A Salt-Caked Smoke Stack

Dirty British coaster with a salt-caked smoke stack,
Butting through the Channel in the mad March days . . .

John Masefield

To the memory of Noelle Pegum
and to our families who had the worry.

A Salt-Caked Smoke Stack

Joan and Robin Markham
and Peggy Lundie

The Pentland Press Limited
Edinburgh • Cambridge • Durham • USA

First published in 1996 by
The Pentland Press Ltd.
1 Hutton Close
South Church
Bishop Auckland
Durham

British Library Cataloguing in Publication Data.
A Catalogue record for this book is available
from the British Library.

ISBN 1 85821 407 6

Typeset by CBS, Felixstowe, Suffolk
Printed and bound by Antony Rowe Ltd., Chippenham

ACKNOWLEDGEMENTS

We would like to thank the following people:
Miss Barbara Bee for professional advice and for typing the manuscript.
Mr Derek Ritson for his support;
Dr Hubert Markham for his rapid co-operation on a weekend;
The contemporary British Consults of Madeira and Las Palmas and the Anglican Vicar of Madeira of that time, for their patient help;
The many people who showed us kindness in Funchal;
The officers of H.M.S. *Duke of York* of the time for their cheerful hospitality;
The Captain and seamen of M.V. *Fort St Joseph* and the Captains of the *Godfrey B Holt*, *Corinthic* and the *City of Exeter*.

PREFACE

The year was 1947, but only just, for the English winter was still at its fiercest. That conflict which our generation still refers to as 'The War' had ended only in August 1945, so the 'peace' was just a little more than two years old. All over the world whole populations were still 'displaced persons'. The unlucky ones had no home to go to. South Africans were luckier. All they needed in most cases was a passage in one of the remaining liners or merchant ships. I am afraid I cannot remember the exact tonnage of the Allies' losses over the preceding six years, but I do remember that one quarter of the peace-time navy had been sunk. Passages to South Africa were therefore as scarce as hen's teeth, and for a South African nurse who had lost her citizenship by marrying an Englishman, the waiting list was a long one indeed.

The waiting period was not cheered by the fact that Great Britain was paying for her victory by poverty and shortages of many things once taken for granted. Even the weather was unkind. It was the worst winter in Europe for many years.

This is the story of four young people who refused to be daunted by all these circumstances. Robin had been through the war as an R.A.F. doctor. Joan had been a military nurse for several years. Noëlle had been a member of the F.A.N.Y.S. Peggy served in the American Red Cross, and later with the Control Commission in Germany. They now tell their story. Noëlle alas cannot do so, for she died some years ago, but Peggy will deputize for her. For obvious reasons the names of the ship and her appalling owner have been changed, but the story is true in every other detail.

FOREWORD

by

Edward Allcard

I first met the Markham family in the Indian Ocean islands of the Seychelles, some twenty years after the bizarre incidents described in this book. I was sailing on my round the world trip. They, in no way deterred by their unpropitious voyage to South Africa, were building a superb cruising ketch for further adventures.

In the ensuing years we have kept in close touch, so you can imagine my interest on receiving the mss of *A Salt-Caked Smokestack* relating events of a story only hinted at during our long friendship. May the launch of this book be as successful as that of their well-travelled ketch *Zamani*.

February 1996

JOAN'S STORY

1. WHY ?

The bed was an enormous edifice of white enamel and imposing brass knobs. On that bleak November morning I lay in it shivering with flu. Patches of mildew formed insistent patterns of geographical fantasy, faces and animals out of the blue and mauve flowers on the wallpaper. A small fire of rationed coal did little to warm the big room. In her cot our well-behaved baby played happily. 'Try to give baby a separate room,' say the books. Sheila had had nine addresses in the first year of her life and was not fussy. She never got my flu. Robin had made her breakfast, taken my temperature, and gone down to the village to telephone. I lay in the virtuous bed and wondered what was going to happen to us.

We were waiting for passages to South Africa, and Robin was temporary assistant to a young doctor of his own year who had built up a flourishing practice during the war while his contemporaries were away. Robin had given up a weekend of demob leave to help him out of a difficulty and felt justified in asking for a day off while I was ill. He worked six and a half days a week for forty pounds a month, with no accommodation and using his own car.

I heard a heavy tread on the stairs, and Robin stalked in. He came over to the bed and looked at me over the brass battlement.

'We've had it,' he said. 'He told me to find another job.'

I said, 'Well, it's been coming for a long time. You've been sweated labour for the bastard – but that doesn't make it any easier.'

We had learned that no doctors wanted a temporary assistant who was waiting to emigrate. There were plenty of 'assistantships with view' – but the view we took of the looming National Health Service was such that we were not prepared to consider partnerships in England. The press was full of vitriolic articles and letters about it all. It was even suggested that people should be 'directed' to areas where jobs needed filling. 'If a dustman can be directed,' asked a Labour paper, 'why should a doctor not be too?' To which a doctor replied that when the dustman could do his job as well as he could do the dustman's there would be no argument.

With so many doctors coming out of the services and the unsettled future of the profession, it was a difficult time. Robin had tried for locums, but these never offered accommodation for a wife and family. If we left our present place we had little chance of finding anything else. We were landed in a huge house, miles from anywhere, with no job – or Robin could find a job, perhaps, with nowhere for us to live. I dreaded the thought of grass-widowhood in that house with its fourteen empty bedrooms, its spooky attic, stone-flagged kitchen, and fiendish pump in the scullery. This pump had an immigrant 'tokolosh' in it, which delighted in squirting icy water from unexpected places. If it scored a hit on one's shivering person its day was made, and it chuckled to itself. You could hear it.

It was all so dismal that I felt that lifting of the spirit that comes when the odds pile up to the point where something is bound to break one way or another. Something has to happen – better the devil you don't know, and any change is a tonic. A foolish feeling, natural enough when one is young. I am afraid I still get it. I said, 'Well, dear, here we go again. The dice are back in the box.'

Robin sat down on the white bed. 'Yes,' he said, 'we'll have to shake it again.'

We had been round all the shipping offices and even to the Ministry of Transport, so quiet and peaceful among the nightingales in Berkeley Square. We were told there were seventy-five thousand people ahead of us waiting to emigrate. I had been to South Africa House demanding repatriation as an ex-servicewoman. South Africa House disclaimed me as an Englishwoman by marriage and was unmoved by my indignant comment on German and Italian women getting priority passages to my country because they had married South African servicemen. At least my spouse had been on our side during the war. It was no good. As the law then stood I had lost my citizenship and rights. It is good to know that my subsequent adventures were eventually cited in Parliament as an example of the need to change it.

Robin had a job waiting for him in South Africa but could get no passage without 'priority' – no priority without a contract – no contract for a doctor unless registered in South Africa – and no registration without domicile. This was a good joke to the clerks in Berkeley Square. We had discussed the idea of flying, but the regular flights were as booked up as the sea routes. Several enterprising characters had bought old Dakotas and filled them with people like ourselves anxious to start a new life. Too many had crashed. There was something wrong somewhere. The DC3 was a reliable aircraft. Anyway they were all charging more than we could afford. As my young brother wrote from his prep school: 'All seems black and ominous.'

There were enterprising gentlemen at sea as well as in the air, advertising passages to South Africa in small craft of various types and degrees of seaworthiness, duly registered as yachts. All aboard shared expenses and signed on as crew. At the other end the owner sold the vessel for a fabulous profit, and returned to England for another ship and another batch of hopefuls. We answered several of these advertisements and got various responses. Usually we were too late. By this time we knew there would be another mouth on our ration strength in about eight months' time. I began to crave for avocado pears, and the South African sunshine became an *idée fixe*. Robin answered the next advertisement by telegram.

The answer came on Christmas morning. 'A wonderful omen,' I said. It was a straightforward letter warning us that the voyage would be no picnic, but if we all pulled our weight . . . etc. etc. Robin was to be the ship's doctor and radio operator. I would be a stewardess. It asked for £130 each for expenses and requested Robin to go to London for an interview.

Our desirable residence stood halfway between Carlisle and Newcastle-upon-Tyne. Hadrian's Wall ran through the grounds. The old house had one 'mod. con.' – electricity – a real anachronism. By waiting until about 1 a.m. I was able to get enough current to defeat 'Shinwell's Cold Snap' (power cuts), and bake some pasties for Robin's journey. His train left for Carlisle and before he left the lights went out. My large husband tubbed himself in the baby's bath by firelight and we dined beneath strings of nappies strung from the picture rails above. I was not happy about my private indoor blackout. Burglars and tramps are all very well, but spooks, no. Even Betsy, a .38 revolver which had bolstered my morale on lonely occasions in Egypt, would avail nothing against 'ghoulies and ghosties and long-leggety beasties, and things that go bump in the night'. That house gave me the heeby-jeebies.

Robin departed into the night pushing Sheila's pram, a great maroon monster which he was taking to his brother whose second child was expected. I would have given much to see his tall figure in a dreadful demob suit, pushing that whacking great pram from Euston to Waterloo, which is what he eventually had to do. After putting it into the Southampton train Robin kept his appointment with the owner of the 'yacht'.

Mr Corbeille was a charming man in his early forties. True, he had a disconcerting facial tic and rather staring eyes, but his manner was perfect and nobody was looking for faults. Everybody in his ship was anxious to get away for various reasons. Very various indeed, as we discovered later. Steam yacht *Woodbine* was a trawler, about twenty years old. She had seen Naval service as a fleet water carrier during the war, crewed by some of our more volatile allies. She had two certificated mates and three engineers. Some of her stokers were professionals too. Robin returned reasonably satisfied and minus £260. He drew me a sketch of *Woodbine* and we started to pack.

How we had acquired enough worldly goods in two and a half years of homeless life to fill fourteen cases remains a mystery. Most of them contained books and liberated radio gear.

We were to join the ship at Southampton. The original arrangement was for *Woodbine* to sail from Plymouth in convoy with another vessel belonging to Corbeille's partner. The second vessel never sailed – for the good reason that her owner was in gaol. We learned afterwards that he had pulled a confidence trick on some incredibly green countrymen of mine while got up as a Captain, R.N. He subsequently married money, rank and beauty and retired to a grand estate, which shall, like him, be nameless.

2. JEMIMA

Jemima was a Baby Austin Ruby Saloon – a hand-painted antique. I painted her myself, so it was genuine all right. When the Royal Air Force dispensed with Robin's profound knowledge of Iraqi hygiene we had set about reconditioning Jemima. She had stood in a broken-down barn, exposed to at least one of the four winds for the duration of the war, providing a comfortable perch for chickens. While Robin attended to her works I removed valuable manure from her person and chipped off most of the rust and old Duco. It took me two days to paint her all over with dark blue synthetic paint, which was all we could get. On the morning of the second day I looked upon the finished half with pride. It was a dreary colour, but it gleamed brightly as I took up the brush to finish the job. I made a stroke – and stopped. I stirred the paint again and made another. My yell brought Robin's head from under the bonnet like a tortoise from its shell.

'Robin! Look at this bloody paint!' Against the shiny dark body two ghastly streaks of sky blue showed up like scars.

'What have you done to it?' he demanded with masculine assumption.

Some time later we discovered that the wretched stuff went like that if one left a partly empty tin, though tightly closed, overnight. Fortunately we had several small tins so I was able to finish the job. Jemima looked quite smart in her shiny blue coat with rusty chromium fittings picked out in black enamel.

At last the great moment came when we were all dressed up to drive to Newcastle and astonish our friends. Robin trundled Jemima out of the barn, while his mother pinned a spray of lily-o'-the-valley on my army blanket coat. Poor Jemima! She was down by the stern like a dog about to scratch its ear with a hind foot. She had broken a spring.

I said, 'I remember an uncle of mine making a wooden block when a spring went . . .' Robin was off. About an hour later Jemima began her twenty-mile limp into Newcastle, with her starboard quarter shored up on a 'gnarly great' wooden block. The cobblestones of Scotswood Road nearly shook our teeth loose but she got there.

Our friends were concocting home-made crème de menthe when we arrived. They admired Jemima and we left her outside their gate and went in to help them. They were mixing gin, peppermint, green dye and golden syrup. It was perfectly horrible, but Tony's determination to see things through had been developed in submarines and was of a high order.

'Just a matter of finding the right proportions,' he said happily, and we tried very hard to find them for him. Finally somebody declared that gin was much nicer without all the gubbins, and that too had to be put to the test. Then Marjory and Tony walked with us down to their gate – and even our hardy submariner was stunned into silence.

Jemima was covered with pea-sized pale blue spots like some surrealist skin disease. We stood goofing at her and looking speculatively at each other. Finally Tony said firmly, 'That car is covered with pale blue spots,' and we all grinned weakly.

'Thank goodness,' said Robin, 'I thought the golden syrup was doing it – or possibly the peppermint essence.'

'It's rain,' I said, 'there was a shower and every drop that lay on Jemima has made a pale blue spot.'

'Interesting sort of paint you used,' said Tony, brightening at the thought. 'I wonder if the spots will stay when she dries off. Perhaps they'll come back every time she gets rain on her.'

Fortunately they did disappear and they never came back, but Tony was right – it was interesting paint.

Now the day came when Jemima was to take us down the length of England. We lashed two suitcases on the back and painted a cardboard number plate to fasten over them. On the back seat went that boon to young parents of the post-war 'population bulge' – a carry-cot. In this went Sheila, hugely interested as usual. On one side went another case, carefully padded with nappies, and on the other one of those shapeless objects known as a parachute bag – I never knew why. What space remained was filled with tins of dried milk, odd items to give to friends en route, and all the things that come under the heading 'last minute'. Before leaving I made a reckless blaze of the last of our fuel and threw some rubbish on top. This set the chimney on fire. Robin put it out.

After lunch we took a thankful farewell of our Bleak House and drove to the village garage – to discover the petrol coupons missing. Eventually we found them in the bottom of a case and set off, once more on speaking terms, at four-thirty. We spent the next day in Newcastle saying goodbye to family and friends, including the crème de menthe fancier, who told us we were several kinds of fools and he was not a bit happy about our 'yacht'. We agreed with him and he agreed with us that we had precious little choice.

Next day thick fog enveloped the whole country, disorganising all road transport. It was clear when we left Newcastle though and parting was a bigger wrench than I had expected. Robin was silent and I felt sad too. I loved Robin's county with its hills and moors. Its spaciousness appealed to me and the friendly people were easy to like. My paternal ancestors were borderers, on the other side, and the turbulent history fascinated me. Robin had shown me Hadrian's Wall from end to end, and we had seen a moonbow over the old military road. Most of my life I have been a person 'of no fixed abode', picking up friendships, education and interests as we went along. Robin's background was similar, for both our fathers were naval officers, though his mother did not 'pack and follow' as mine had done. For me and my sister Barbara every three years brought a change of schools and familiar things. We were born with wanderlust but never get used to the partings. However it is no good hanging over the stern getting sentimental, especially as you know you would be furious if you were suddenly whisked back. Better get along forrard and look ahead.

Jemima was unheated. Icy wind whistled in at every crevice and while I attended to Sheila Robin took our ration books out of his briefcase and bought me some woollen stockings. He put the books into a pocket of his ginger plus-fours, and this saved our bacon, literally, later on. (The initial shock of ginger plus-fours to a colonial bride who had only seen her husband in uniform had not worn off. It never did. He and his friends dug out these weird pre-war garments and the hairy objects never wore out. Even the ubiquitous skimpy demob suits were less ugly – but Robin's festoons of ginger tweed did keep him warm in Jemima.)

After a good lunch at Scotch Corner we set off down Leeming Lane. Then we remembered Briefcase. It had been dumped on the over-laden luggage rack while we stowed Sheila and neither of us had put it back in the car. It was gone, of course.

Briefcase was a present I had bought for Robin at Simon Arzt in Port Said with the last money I earned before Sheila. I had braved the Canal Road from Ismailia specially to buy this unfortunate present. I had chosen Briefcase after much thought to counteract our gift for losing things. I had put all our papers, my jewellery, even my watch (broken) and engagement ring, savings certificates, insurance policies – and our passports – into Briefcase, saying to Robin: 'Now there's only one really important item and it's easy to watch.'

The ship was supposed to sail in a few days. I remembered something in my passport to the effect that, if lost, it could only be renewed after 'exhaustive enquiries' had been made. We had passed a convoy of tanks on the road and I could almost feel them crunching over my jewellery. We turned back. There were about forty men working on the road and we made exhaustive enquiries of every one and of the hotel where we had lightheartedly enjoyed our lunch. No Briefcase. Frantically we hunted for a police station and finally found one at Londonderry in Yorkshire. The sergeant's wife took me in and made tea while Robin talked to her husband. The typical north-country kindness touched me very much. I had a weep and promised to tell her the end of the story.

There was nothing to do but press on. The fog closed down and visibility was reduced to nil. By nightfall Robin was very tired, but thinking of those exhaustive enquiries we thought we had better get to London as soon as possible. I had no driving licence in those days so he had to do it all. Our lights were useless in the fog. Robin drove very slowly while I hung out of the open door and watched the side of the road which was invisible at three feet. After several hours we found a big lorry that had run over the edge of the road. Robin helped the driver to get it back and lent him our torch. He told us to creep along behind him as his fog lights were better. We did this thankfully and presently we realised we were part of a long convoy. Jemima must have looked funny in the middle of that procession of giants. The strain was terrific. We were about two feet behind our lorry and Robin had to watch his brake lights and continually change gear. We began to worry about petrol.

We crawled into Doncaster, and two hours later we crawled out again, having seen nothing but the rear lights of our lorry and an occasional glimpse of kerb under a patch of paler fog which indicated a street light. After a while we decided to call it a day. Somehow we found a hotel in Bawtry. I went straight to sleep in spite of the cold and woke to find Robin pacing the floor. Next morning the fog showed no sign of lifting but we set off once more with visibility nil. After a few hours it got patchy and we ran into clear daylight for several miles. At once things seemed more hopeful. We felt almost cheerful and very hungry. We had lost our lorry during the night but stopped at a lorry drivers' pull-up and had dehydrated scrambled egg on toast. As usual Sheila broke the ice and soon the whole crowd was sympathising over our loss. The proprietress of the café said she would ask every driver coming through if he had heard of anyone finding a briefcase. Sheila made instant friends with a very attractive girl – an unexpected vision of smart clothes and good grooming. It seemed centuries since I had looked like that! She had a friend who had lost his passport that same year. 'What's the form?' we asked her. 'What about all those exhaustive enquiries?'

'There's no enquiry at all,' she assured us. 'You just breeze into the head passport office and tell your tale. You'll get your ship all right.' Clouds of anxiety rolled away as the fog had done. She must be about seventy now but still remembered as a beautiful girl in a café on a fog-bound road.

At midnight we drew up at our old digs in Farnborough, where we had lived when Robin was S.M.O. of the R.A.F. station. My old grandfather was staying there now. His wife had died during the war, his family had settled overseas, and when we arrived in England he was living in appalling rooms with a landlady who pinched his rations. When we left Farnborough we asked our dear old landlady to take him in. He had been on the waiting list for a passage to South Africa since 1940. It was granted about six months after we left and he died two days before his ship sailed.

Leaving Sheila with our beloved landlady and Grandfather, we went to London, got new passports with no trouble at all, and drove back that night with snow falling and Jemima's tiny radiator covered with cardboard. The following morning we took Sheila up with us for yellow fever jabs and more farewells. It was freezing hard but she enjoyed the changing scene from her carry-cot, shouting remarks to strangers who invariably stopped to wave. Next day we left for Southampton, leaving the tall, fine old figure of my grandfather waving goodbye to us in the falling snow.

3. SOUTHAMPTON

Snow was lying a foot thick on the road but the air was crisp and clear. We bought a tin of sausages 'on points' and ate them in the car. We stopped at an antique shop and bought a pewter mug for Robin's collection – the second we had found on the journey. They cost about ten shillings; fifty of the regrettably named 'pee'. Southampton looked fresh and clean under the snow. We drove to Robin's brother's house and relaxed thankfully by the fire. It was good to see them again. My sister-in-law and namesake was a beautiful woman who worked all the hours of the day. She was the perfect doctor's wife. Hubert's practice was busy enough without the innumerable forms and paperwork that were already an increasing burden on the profession. We agreed that we would face almost anything rather than accept a life like theirs.

We had long arguments about emigration. Hubert and Joan felt it was a duty to remain and face things out. We argued that calm acceptance of circumstances never got people or nations anywhere. Between us we represented the two seemingly conflicting types who between them produced the Empire – the individualists seeking individual freedom with, if possible, a bit of adventure thrown in, and the sound, worthy citizens who can live in a dockyard town without the slightest interest in the ships – the 'English of the Island', indestructible and enduring, and impervious to their climate.

Two days after our arrival in Southampton a policeman on a motorbike came to the door – with Briefcase! A lorry driver had handed it in to a police station and they had traced us in less than a week.

Everything was inside, undisturbed. Robin got the driver's name and address from the policeman and sent him a reward. He got a nice letter back saying they had bought a carpet for their house with it, and he was glad to have helped. I wrote to the sergeant's wife in Yorkshire and told them the outcome of the story as I had promised. 'This is another good omen,' I told Robin, looking at my engagement ring which I had thought I would never see again.

* * *

Robin skirted a bomb hole on the snow-covered wharf and parked Jemima. I looked at S.Y. *Woodbine* with mixed feelings – fear, excitement, and a totally irrational immediate liking. I think the fear predominated. I said: 'Well, this is the first time I've walked *down* a gangway to board a ship.'

A dazzling figure stood on the diminutive deck. One had a swift mental picture of a capital ship of the home fleet before a Royal inspection. Four gold rings flashed from the cuffs of an immaculate monkey jacket. Gold gleamed upon a peaked cap worn at an Admiral Beatty angle. All this glory was strutting around the dirtiest little ship we had ever seen. We went on board. Trampled snow lay on the deck mixed with several weeks' accumulation of galley refuse waiting to be thrown over the side outside Southampton Water. She was still wearing her wartime grey paint, heavily stained with rust and dirt.

Corbeille bore down on us with pompous geniality. We gazed upon the uniform in fascination. The four gold rings were surmounted by a gold 'diamond' – an intriguing compromise between the Royal and Merchant navies. The cap badge was that of a famous yacht club. We were hustled along to meet his newly acquired wife.

In my old khaki skirt and woollen stockings I was introduced to a trim little person who could have posed for one of Hynes glamour girls of the time. She took full stock of my practical but unattractive dress, from the woollen turban to the thick service shoes, and I saw her mentally docketing me as harmless. She made some tea and confided to me that she was in a difficult position.

'You see, I'd like to – er – muck in with the rest of you, but I have to remember that I am the Captain's wife.' I managed to keep a straight face.

Corbeille and Robin returned from an inspection of the ship's radio. 'You may live ashore,' he said, 'but I expect everybody to work preparing the ship and loading stores.'

There was plenty of work. I was handicapped by Sheila and could not always get down to the ship. Corbeille was gallant about this. 'You just do what you can,' he boomed. 'We all understand about the baby.' He was certainly getting plenty of work out of Robin. He had bought an old R.A.F. 1154 transmitter which was completely unserviceable. Robin had no diagram of its interior but he took it to the local R.A.F. station and worked on it till late at night. He climbed the single mast and the tall thin funnel to fix the aerial. He installed his own receiver. He made a barricade of wooden slats to keep Sheila from escaping from my cabin. This had been the radio officer's office. It measured six feet by five. The big operating desk was still in position and Robin built me a bunk a foot above it. He railed suitcases into the space between. When I lay on my bunk my face was level with a six-inch porthole facing forward. I liked this. The parachute bag and baby's bath went in the knee hole under the desk and the two small cupboards flanking that held tins of baby food and the nappies. This left just enough deck space to wedge in the carry-cot and for one adult to stand up. I scrubbed this 'state room' with disinfectant and stowed the gear.

Other members of the crew were joining daily. I was peeling potatoes in a bucket of cold water and trying to ignore my chilblains when the cook stuck his head in the galley and said, 'The other women have turned up.' I hoped they would be good types but did not expect too much as the rest of the crew seemed to be a fine collection of eccentrics.

On the quay above us a group of amateur sailors was hauling on a thick and filthy line. A slim blonde girl in slacks was pulling away with the best. Oh dear, I thought, a hearty type. A second girl was standing by one of the two forecastle doors. I walked over and introduced myself to a tall cheerful girl with beautiful hair and a real English complexion, dressed in a most unbecoming pair of baggy corduroy slacks. This was my first impression of Peggy, who is now a godmother of our second daughter. She told me the girl on the quay was her great friend and that they had just returned from Germany, where Peggy had been in the Control Commission. The blonde came aboard and joined us. I liked Noëlle at once – an Irish type with a refreshing lack of self-consciousness and a glint in her eye. Not 'hearty' at all. Robin arrived, met the girls, and an alliance began which has lasted ever since.

Sheila was to be quite a refining influence when things got bad – an important little person who hit the headlines at fourteen months and accepted the adulation of the crew as her due. When she grew up we enjoyed telling her how a hard-case Greek fireman once presented her with a little chain bracelet in a Spanish café and how an equally tough South African deckhand once stole fourteen bars of chocolate for her.

There were several paid hands among the twenty-seven of us, as well as certificated mates and engineers. We realised later that men holding tickets were hardly likely to sail in a vessel like *Woodbine* unless they had urgent private reasons. They had. The Chief Engineer, a stout sentimental man with a glass eye, had burned his fingers in a second-hand car racket. The Second turned out to be a dipsomaniac with a forged ticket. The Third was a decent youngster who seemed to be rather under the thumb of the Chief, who lived next door to him in Belfast. The Mate was always called Skipper, which indeed he was, for Corbeille was only the owner, of purely decorative value. The Mate was a good seaman and navigator, but had a painful inferiority complex which sometimes made him rude and aggressive. His wife was expecting a baby in South Africa.

The Second Mate was a little old Geordie, who, in his seventies, had survived several hours in the sea off Newfoundland after his ship was torpedoed. He looked a bit like Popeye. He loved to spin long and actually very interesting yarns in a slow drawl which maddened his audience. Dentists had obviously meant nothing to him in his long life. He had his wife with him. Mrs Hoppy was a cockney with a sublime intolerance of all things foreign. She had little hair on top, a kind heart, an acid tongue, and a glass eye (on the opposite side to the Chief's). She and the girls shared a section of the forecastle, partitioned off from the men's quarters. It had almost no ventilation except through the door, and was dark even in daytime.

Corbeille missed no opportunity to make money, and he had shipped an extra boat to sell in some port of call. This was much larger than Woodbine's own boats and was stowed forward on the well deck. Beside it was an enormous hemp fender, big enough for one of the *Queens*, curled up like a roly-poly. When this monster was shipped Noëlle's comments on 'that bloody great pudding' were pungent. It meant that their door could only be opened a third of its full extent. Should the pudding shift eighteen inches they would be penned in their little black hole – not a happy thought later on, in a full gale with no way on the ship.

Joan and Hubert's second daughter was born a week before we sailed. Sheila and her elder cousin played in the house, supervised by a very superior maid, so I was still going down to the ship in the mornings, peeling potatoes and washing dishes with Peggy and Noëlle. I asked Mrs Corbeille why the girls should go into the fo'c'sle after every meal to wash the men's dishes after washing up for the officers aft.

'Well,' she said, 'they signed on as stewardesses. I also have to work – as purser.'

I tried to raise an eyebrow. 'Since when have stewardesses waited on the fo'c'sle?'

The Chief butted in. He loved to play the role of protector of helpless womanhood. 'It's disgraceful,' he said unctuously, 'nice girls washing up for a lot of common sailors.' I fled.

The girls stopped going to the fo'c'sle. I had helped them on one of these occasions and had noticed that while the men showed me a certain amount of respect, their attitude to the girls was a bit off. I was rather puzzled but put it down to the fact that I had a husband on board. Some time afterwards one of the men explained it to me. Corbeille had told the crew: 'I'm taking a couple of women along, so you'll be able to get a bit of fun there.'

One of the boys, aged nineteen, known to all as 'the Gentleman Spiv', was expecting his wife to join him. We turned up one day to find Peggy and Noëlle rolling on their bunks with laughter. The bride had arrived – 'petite blonde, with a hairdo just like yours, Joan.' She had walked round the ship turning up her respectable little suburban nose at the squalor and gone home. The girls had heard her parting remark to her husband. 'I could not live with those common women.' The story took a little while to tell because the common women kept collapsing with mirth.

Sailing was delayed so often that morale was sagging. One of the crew, having successfully sought consolation ashore, went peacefully to sleep in the tiny 'wardroom' amidships. One of the officers ordered him forward. He argued. The officer pushed him out on deck, with no ill feeling on either side. Corbeille heard the row, dashed down from his cabin in pyjamas, and began to beat up the dozy drunk with extraordinary ferocity. Shouting: 'I am the Captain of this ship!' he kicked the man ashore and threw his gear after him. A dockyard policeman collected the lot and we saw him no more. The incident caused a good deal of unfavourable comment on Corbeille's methods, but we did not yet know that battle cry: 'I am the Captain of this ship!' and the excitement died down.

There was a further delay when the Board of Trade refused to pass our lifeboats. This was hardly a surprise as they were several sizes too big for the ship and could not be launched at all. Corbeille conceived the brilliant idea of sawing the davits through and welding them together again – presumably to let a piece in? The Skipper managed to squash this idea and the davits were unshipped and remounted. The boats were duly lowered before a Board of Trade official. The operation took three-quarters of an hour each, in harbour, but the official seemed satisfied . . .

The stores arrived and all hands were put to loading them. This was difficult for me. The other girls were heaving cases of whisky aboard (with enthusiasm). I was getting hard looks from some of the crew for slacking, so I confessed our little secret to the bosun and after that they looked after me like a lot of mother hens.

The bonded stores set a major problem. They had of course to be locked up and sealed until we got to sea. There was an amazing stock of wine and spirits. Only the most optimistic believed they were all meant for Corbeille's hard-working crew, but you could feel the morale bracing as each case came aboard. The problem was simple. There was nowhere to lock them up. A brilliant solution was finally reached. The bonded stores were locked in the fo'c'sle lavatory and the door was sealed with due care. Next morning we arrived to find a tremendous commotion on board. The seal had been broken during the night and a case of whisky was missing.

Corbeille was incoherent with rage. We gathered that it was going to cost us about forty pounds each and he was the Captain of the ship. Of course nobody had the slightest intention of forking out forty pounds. The air of mystery deepened and was hugely enjoyed by everybody – until the missing case turned up in the ordinary storeroom where it had been all the time. Finally a Dutchman called Joe confessed that he had broken the seal. 'I do not know the bloddy head is vol whisky,' he protested in an injured tone. 'I am ashore met my girl – I do not know you put bloddy whisky in the bloddy shithouse – is dam-silly idea anyway.'

Joe was a character one would prefer not to meet on a dark night. He made one think of razors. The consensus was that he would cut a throat without changing his expression. He took infinite pains with his appearance and had a very natty uniform of his own devising. The men called him 'the Chinese Admiral'. He was always ashore with his girl, who was what you might expect only more so.

Going ashore from *Woodbine* required agility and nerve. At high tide the boat deck was just above the quay, and at low water a good fifteen feet below. A ladder balanced on the well-deck formed our precarious link with the shore and the ordeal of creeping up or down this over a strip of icy harbour water can only be appreciated by people who share my horror of heights. At last we managed to improve the situation. While walking along the jetty one dark afternoon Robin and I saw a gangway – a beautiful gangway, all by itself, with no line or company mark on it. Hastily fetching Jemima we hitched her up to the treasure. Robin drove slowly and I walked behind guiding it like a plough. Suddenly, a voice: 'Wot's all this 'ere?' – and there was a burly dockyard policeman. 'Wot do you think you are doing?' he demanded patiently.

'Oh, well, er – just pinching this gangway,' Robin answered, thinking, 'This is the end—!'

'You come from that old trawler over there?'

'Oh – er, yes actually . . .'

'Well I reckon you need this gangway pretty badly!'

And the Law obligingly helped us drag our prize along to the ship.

Every day Jemima waited beside the bomb hole in the quay, her little radiator drained and her bonnet padded with straw. Snow and sleet were churned to filthy slush. It was freezing hard and men were oiling the points on the railway lines on the quay. A few hundred yards away one of the *Queens* lay at her berth, forming one of the contrasts which were to be so frequent in our near future. Thinking of her pampered passengers we reminded each other that we too had paid for our berths – and the privilege of working like galley slaves in our little tub. Naturally we felt superior in our discomfort – an utterly unjustified but very human feeling. We could still laugh at ourselves.

Still our sailing was delayed. Suddenly Corbeille got an urgent call to London. He came back in twenty-four hours and immediately sacks of coal arrived and were loaded aboard in an urgent rush. We were sitting in the mess aft, a small space next to the engine room, lined with bunks and furnished with a diagonal table, benches and a smoky stove. Above the stove the dishtowels hung amongst the Skipper's vest, the Mate's socks and the Engineers' overalls. Beneath this collection stood a couple of buckets of greasy water for washing dishes. One had to be careful when reaching for a dishcloth to grab the right article.

The Chief came down the iron ladder at the foot of the table. He was large and the hatch was small. One of our amusements was to sit at the table and watch people descending. It was an unforgettable sight to see the Steward come down. He was a grossly fat man with an enormous posterior and he always came down backwards. As his vast stern hove in sight conversation stopped. He always just made it and would subside upon the nearest seat and begin highly coloured reminiscences which began: 'When I was in San Domingo . . .'

This time the Chief was upset. 'The bloody fool is loading household coal,' he snorted. 'We ought to wait our turn for steam coal. I told him I won't be responsible. This stuff will burn up in a few puffs of smoke.' The Skipper pricked up his ears.

The ladder began to shake again and Corbeille's elegant figure appeared above us like a modern annunciation scene.

'We are sailing tomorrow. You will all go to the Emigration Office in the morning and report on board at eleven.'

The Chief said, 'That's impossible, Mr Corbeille. This ship has been laid up for over a year and we haven't had a trial—'

Robin said, 'I've not been able to test the radio yet.'

Corbeille scowled. 'You will address me as *Captain* Corbeille. I am the Master of this ship and on board my ship you will do as I say. We will sail tomorrow.'

We spent the evening with Hubert and a cousin of theirs, a Merchant Navy captain of a salvage vessel. He was lifting the pipeline 'Pluto' that had carried oil across the Channel for the Normandy invasion. He had been aboard *Woodbine* several times, and had shaken his head at the extra boat and the pudding-fender. He had given me a roll of bedding and shown me the correct way to make up a bunk. I wished so much that he and his inseparable little dog could sail with us. Once I actually said to him impulsively, 'Come with us, Tom. I'm scared stiff.'

He laughed. 'The *ship's* all right,' he said. 'I wouldn't mind taking her to Cape Town.' He had to take his own ship out next day so we did not see him again, but I can picture him now, standing on *Woodbine's* fo'c'sle shaking his head at the 'pudding', with his little dog looking just as concerned as his master.

4. ON PASSAGE TOWARDS FUNCHAL

After so many delays Corbeille's sudden insistence on hasty departure took *Woodbine*'s people by surprise. We arrived on board a little late and were soundly reprimanded. Sacks of coal were stacked around the deck on the cinder-powdered snow covering the accumulated gash. Corbeille was fussing about in a most unseamanlike manner, getting in the Skipper's way and reminding everybody who was Master of the ship. The amateur sailors were falling over each other in true Oakley Beutler style. The inevitable humorist was shouting, 'Any more for the *Skylark*?' and getting sworn at by the Skipper.

Finally the moment came. 'Let go forrard!' The Gentleman Spiv made a spectacular leap over the fo'c'sle head. On the bridge Corbeille gripped the rail in war film style – Captain D taking the flotilla into action (the authenticity of the effect spoiled by shiny gold rings well to the fore).

Beside me on the narrow deck outside my cabin Robin stood holding Sheila, separated by four feet from his brother and her little cousin. Jemima stood forlornly behind them. They were the only people on the quay. I thought of pre-war voyages that began with bands and paper streamers – scenes from another life. As we drew away slowly our last memory of England was the picture of the long thin figure of Hubert in his old Burberry and his little daughter alone on the quay in the wind and snow, with our little blue car behind them.

Woodbine was making sudden sorties in all directions, swinging the compass while still in Southampton Water, while Corbeille argued with the pilot, who was horrified at the idea of sailing without trials. I had been almost immune to seasickness all my life but in view of pregnancy and the weather I determined to get Sheila fixed up before we turned the corner and the fun started. I took her down to the mess aft, bathed and fed her and handed her over to Mrs Hoppy while I hastily dealt with the nappies. (No disposables or detergents then, only yellow soap acquired on the black market.) I was in the engine room, on the catwalk above the engine, hanging them up on a line rigged for the purpose by the Chief Engineer, when Corbeille roared through the door: 'What are you doing in my engine room? Get out at once!'

It was too much. 'Mr Corbeille, I will get out of the Chief's engine room when he tells me to.' He went on shouting but I had turned back to the nappies. The Chief lumbered up from below. Young motherhood being bullied, *and* his authority being usurped in his engine room! I left them to it and joined the other 'idlers' in the mess. They were drinking hot lemon juice and persuaded me to have some, just before we turned down channel – against the tide. What a mistake! I fled on deck. The wind caught me and I stumbled hard against the low rail. That was the end of the lemon and nearly of me but Peggy had grabbed my ankles as I lay across the rail and was hanging on. We had about eighteen inches of freeboard and were rolling hard. I got back inboard on the opposite roll.

I refused to let anyone carry Sheila up to my cabin. She spent the night in the mess with Robin and the other officers. Robin, who is never seasick, tested the radio and to his relief it worked. I don't think anyone on board realised how much time and work he had spent on it. Fortunately for all of us he is a first-class operator, known in amateur circles all over the world, with a morse speed of forty words a minute, and an expert knowledge of radio in general. He worked Niton Radio and later the Lizard. That night he joined the 'gang' on eighty metres and arranged with his friend Derek for nightly contact at 10 p.m.

Next morning the weather was better. The swell seemed tremendous but we knew that a big ship would have been feeling no more than an ordinary roll. For the first time in my life I experienced the horror of seasickness – not the kind that wears off in a few days but the real thing. I made myself go out on deck. Sheila was asleep below and the girls were peeling potatoes outside the galley. The average sailor – and soldier – would insist on having the spuds peeled if the world collapsed around him. I took a knife and started to help. They told me to go and lie down and not be an idiot, but my pride was suffering. 'If I can only keep going I'll be all right,' I said, and thought bracing thoughts. A fine thing if your piratical ancestors had gone to lie down every time they felt seasick. Just look at the spuds and keep your eyes inboard.

The peeling was finished without mishap. The girls went below to wash the breakfast dishes and I followed them. A blast of hot engine-room air nearly finished me but I passed that obstacle safely and descended the famous ladder. The indescribable atmosphere of food and frowst that greeted me was too much. Half an hour later I lay shivering on my bunk with Sheila playing in her carry-cot on the small deck space below me. 'The piratical ancestors weren't expecting babies,' I consoled myself in my misery. That was my last attempt to go on deck for fourteen days. I got weaker and weaker and began to fear for the new baby. Several of the crew were sick too, but Peggy and Noëlle were carrying on. The galley of course was a shambles and one day the cook's false teeth turned up in a batch of scones. The girls made rock cakes in the coal stove and we bowled along at a steady eight knots.

Robin stood his watch at the wheel, washed nappies, fed Sheila and attended to the radio. I fretted in helpless misery, feeling utterly useless. I could not keep even a sip of water down. After a few days Robin was giving me injections to prevent the loss of the baby.

ROBIN WRITES:

Like all the able 'passengers' of the crew I was on the wheel roster. My watches always seemed to cover the small hours and I could see nothing at all of the sea or the horizon – or for that matter of the stars since the weather was nearly always overcast.

The bridge was a large, almost empty compartment on top of the ex-radio room and the cabin of Corbeille and his wife, with round-the-compass views out into the Atlantic darkness. The wheel was a huge spoked contraption about six feet in diameter slightly forrard of the centre of the bridge space, and overhead, suspended from the deckhead, was the compass with its dial and needle visible from below. The helmsman could check his course by glancing upwards. There were flag lockers on the bulkheads and a chart table. In the centre of the deckhead was a single electric light fed by the ship's 110 volt D.C. supply. The light was dim and had the characteristic red tinge of an under-run bulb.

Most of the time during my watches I stood up there alone, swaying from one foot to the other as *Woodbine* rolled. One night, somewhere off the coast of Portugal, Corbeille appeared in full panoply of Master of the Ship and stood behind me while checking the course – I never saw the relevant charts. He asked me whether all was well and on receipt of my 'Yes' – (no 'Captain') he began to make conversation by asking whether I had been to South Africa before. I told him I had spent some time in Durban, having been put ashore from a convoy bound for Egypt, and hospitalised with pneumonia. He told me that South Africa was a country of great opportunity, and said he had a property in the Eastern Cape. (I later learned that he had a dispute about his boundary with the owner of the adjacent land.) Then he took from his pocket a wallet and out of this took a glistening stone the size of a small marble with sharp angles on it. It was a light brownish-yellow and I said, 'It looks like a diamond.'

'Yes, it is,' he said, and turned it round and round in his fingers, squinting up at the light coming through it from the deckhead bulb. The yellow hue of the latter deepened the yellow of the stone. He put it back in his wallet and returned it to his breast pocket. I made no comment and didn't ask him how or where he came by it and he volunteered no further information.

Why he revealed to me this probably illegal possession I never found out. Possession of uncut diamonds without a permit is looked upon seriously in South African law. No person who is not a registered dealer or cutter of raw diamonds is allowed to possess them and even the individual stones are enumerated. Perhaps this incident was a kind of test. If this were so I failed it.

I said nothing. I was worried about Joan and our baby lying a few feet below me in the minuscule cabin with its door giving onto a narrow deck that ran forward and then to starboard athwartships, under Joan's port and those of Corbeille's larger cabin where it met the head of the slanting ladder down to the deck.

During this part of the voyage I was passing the open doorway of the engine room where a ladder led down vertically to the engine footplate. I stopped and looked for a minute at the engine. With its two vertical cylinders it stood some eight feet above the exposed crankshaft which was level with the footplate and was turning over about once per second.

Reciprocating engines always look as if they are really busy. Piston rods going up and down, crossheads rocking to and fro, and the big ends cranking up and down under it all. There was a small tank situated near the top of the high pressure cylinder (looking very small beside its low pressure partner which was about the size and shape of a forty-gallon drum). This small tank received freshly produced hot water from the condenser and formed a reserve before being pumped back into the

boiler for the whole process to be repeated. It had a lid – somewhere.
The holes for its holding bolts were there but they were empty. There
was no lid visible. At every roll hot water cascaded down into the bilge
and was, presumably, pumped overboard by the duty engineer when it
got too plentiful.

I remarked about this to the Chief and asked what he would do
when the reserve of fresh water ran out.

'Oh, then you pump in sea water,' he replied, 'and you're running
on your ocean.'

I said, 'Chief, I'm not an engineer, but I'm sure boilers are not meant
to boil salt water – and doesn't salt water take more coal to boil than
fresh?'

He assured me that it wouldn't matter.

JOAN CONTINUES:

We were about a week out, rolling badly in a heavy swell, when I awoke
one night with a raging thirst. I remember a heavy blow and standing
with my back to something, staring at heaving black water, conscious
of a great fear. Then I was lying in the little 'wardroom' – faces looking
at me – Robin – door slamming – an empty bottle rolling to and fro on
the table. A hand took it away.

They had been sitting in there where Robin had installed his radio
gear when the Skipper kicked the door open and staggered in carrying
me. He had found me, completely unconscious, lying across a rung of
the bridge ladder, head and shoulders on one side and legs on the other,
swinging out over the sea with every roll to starboard. Somehow he
got me back and carried me down. Robin took one look, saw I was
alive, and dashed up to see if Sheila was safe. She was in the carry-cot,
the barricade firmly in place and the door shut. The blow I remembered

must have been its slamming on me. The Skipper was not a big man and it must have been a difficult and dangerous feat to get me off the ladder. I would inevitably have slipped right through it into the sea had he not found me quickly. Nobody would have known what had happened.

On the twenty-fourth of February we came in sight of Porto Santo, the northernmost point of Madeira, with its towering mountains and the little off-lying island bearing the Decima Light. I could see the island from the porthole by my face. Thankfully I drifted off to sleep again. I was awakened by a tremendous crashing and banging. I thought, 'Now they've run the ship ashore and we're breaking up,' and passed out again. I was unconscious most of the time now, with intervals of surprising lucidity. I had a series of vivid impressions strung out on the general period of unconsciousness like beads on a string. When I came round again the engine had stopped and the ship was rolling worse than ever. The noise of wind and sea was very loud. Robin came in and I asked, 'Why has the engine stopped? We're still at sea.' He tried to reassure me, but I insisted. 'We're rolling like hell and there's no way on the ship.'

He took my hand and said, 'Darling, we've run out of coal.'

'What was all that noise?'

'They were chopping up the wooden platform and fittings over the steering gear to try and keep steam up to make Funchal. Corbeille even wanted to burn all our baggage! It was no good of course, but I've radioed Madeira and a tug is coming. It will pick us up about 10 p.m.'

Hours and hours to wait, I thought miserably.

Robin was unwrapping a parcel, balancing on the heaving deck. Nothing surprised me now. Sheila sat in her carry-cot, chattering and thoroughly enjoying the delightful 'ride'. Robin said, 'Darling, many happy returns of your birthday,' and showed me the contents of his parcel – a pair of antique pomade jar lids for my collection. In all the rush at Southampton he had remembered to buy a present and now he remembered the day. It was my twenty-fifth.

All night the wind rose until it reached full gale force. We tossed about in seas over thirty feet high. My bunk was athwartships, my pillow had disappeared, and the top of my head was bruised. The bunk was six feet long and I am five foot six so with every roll I slid six inches, alternately crashing my feet and head on the bulkheads. Sheila's cot was well jammed in and she slept most of the time. She never cried once. A box of puffed wheat had fallen on top of her and burst and she sat happily munching and playing with it for days . . .

ROBIN:

The tug never found us. The overlying decking which covered the long tiller of the steering gear supplied about a minute's heating in the furnace as it blazed away up the funnel. The tiller now tugged at the steering chain as the beam seas beat against the rudder. I don't remember whether the engine showed any signs of turning.

After a few minutes of discussion with the Chief Engineer Corbeille told me to send a signal to the Port Authorities at Funchal requesting a tug to meet us at a position twelve nautical miles to the south-east of us. The Skipper protested.

'We can't make it. She's drifting rapidly northwards.'

Corbeille replied, 'I intend to sail the distance.' Knowing there were no sails aboard we ignored this strange statement.

I asked him to write a proper signal and sign it. He said, 'You will send the message as I tell you.' I wanted it in his own words as a signal from the Master. Eventually he agreed and gave me a piece of paper with the message on it. I went and switched the transmitter on and gave the message to Funchal. I arranged a schedule for one hour later and received a reply that the tug *Gaviaio* was setting off, to meet us at approximately 22.30 that night.

Darkness fell as we lay on our bunks and the wind rose steadily and rain fell in torrents. At 22.30 visibility was nil. No tug! If *Gaviaio* had arrived at the arranged rendezvous in clear weather I could have worked her easily on the Aldis lamp, but it was not to be. Had Corbeille's signal given even the approximate position to which we would have drifted there would have been a strong chance of contact. As it was, the prospect of a cheaper tow starting within the territorial waters of Madeira overcame Corbeille's better judgement, if he had any.

The tug *Gaviaio* had no radio.

The 'wardroom' was empty that night. I told the operator in Funchal, Henriques (we were on Christian names by then), that I would work an hourly schedule just as a check on the situation on a frequency in the 500kc shipping band that would not be interrupted by other traffic.

JOAN:

All the next day we drifted away from the island. Many of the crew were sick and the fo'c'sle must have been terrible. In the Black Hole Mrs Hoppy had developed what used to be politely called 'gyppy tummy' – or in hospital circles 'D and V'. She made full use of the enamelled bucket which normally served for their ablutions and then upset it. Collapsing on her bunk she left the girls to clean up the results. They had to live in that atmosphere for the rest of the voyage. The last time I was in the Black Hole, weeks later, it was still apparent.

There was no coal for the galley stove. Peggy and Noëlle served what meals were possible to those of the crew who were able to face them. With no steam for the generator there were no lights. They made tea with a blowlamp. One night the drunken Second Engineer flicked the blowlamp on the seat of the Steward's trousers stretched taut across his enormous behind as he made his usual precarious descent of the mess ladder. The girls said it was one of the highlights of the voyage.

The constant heaving about had turned the water tanks foul. The water was green and smelly and undrinkable. Fortunately it was raining hard. Robin stretched canvas under the bridge ladder and caught rain water. It was sooty and tasted of canvas but we were lucky to have it. He made Sheila's feeds with it, using the famous blowlamp. He was snatching a few minutes' sleep when he could, constantly attending to the radio, taking his watch, washing the nappies – which could not dry properly when the engine stopped – and doing his best to keep me alive. He told me afterwards that another four days could have meant the end for me. I was seriously dehydrated and he was giving me small shots of morphia and pouring some of the sooty water down my throat, slapping my face to make me swallow. He was still giving me injections to prevent miscarriage, which would have finished me as well as the baby.

During the day an American ship passed us within a mile. Fire was lit on the fo'c'sle, distress signals were hoisted. The old rhyme goes:

'When in danger or in doubt
Wave your arms and jump about!'

The crew did that too, but the ship sailed unconcernedly on, – her watch keepers, if any, probably saying: 'The Limeys must be having quite a party.'

ROBIN:

I missed all the fun on the foredeck when the apparently unmanned ship passed us. I was sound asleep after a night of radio vigil. After a disturbed night on watch and radio alert I slept through the drama.

When I awoke and looked out forrard I could not believe my eyes – four brown army blankets were tied by their bunched-up corners to the mast stays. The wind had dropped to a faint breeze and they bulged to leeward but made little difference to our way – which was *away* from Madeira and towards the inhospitable coast of North Africa some three hundred miles to the east.

When I went up to the bridge Corbeille ordered me to send a signal to Funchal informing the Port Authorities that he was sailing in under 'main course, topsails and royals.' I admit I sent the message with a certain relish.

I was extremely worried about Joan. That evening, without reference to Corbeille, I called Henriques at Funchal and asked him to advise neighbouring shipping that we were in need of assistance. The signal was *not* an S.O.S.

During the afternoon of that day I had privately gone up to the bridge and noted the wind direction from the compass. By throwing a half-filled bottle with a plug in the neck into the sea and timing our drift past it downwind I got an idea of our speed of drift – the old 'Dutchman's log'. I told Henriques our approximate position was eighty miles north-east of the Decima Light. He took the signal and arranged a schedule at 22.00. At 22.00 he told me he had advised two ships of our position and plight, and I went to sleep in my bunk next to the engine room door.

At 02.00 I was awakened by one of the young boys in the crew who shook me and said, 'Doc, come onto the bridge – there's a ship coming.' I was fully dressed and went straight up and sure enough there were lights to the *north* about five or six miles away. I picked up the Aldis and he answered straight away.

'Are you *Woodbine*?'

'Yes,' I replied. He gave his ship's name as *Godfrey B. Holt*. Meanwhile Corbeille had been called and appeared at my elbow. The ship flashed,

'What is the nature of your distress?' I replied that we were out of coal and had a potential hospital case on board. Corbeille then told me to send a signal asking if they would transfer some coal! Although the wind had dropped there was a heavy swell and so the idea was ridiculous. What the Captain thought of it I could imagine. His negative reply was not surprising.

While this was going on one of the lads, who were naturally on the 'monkey island', called 'Doc – look: there's another ship!' Sure enough, there were the lights of a ship approaching from the *east*. The two ships converging at one point in the ocean could only have meant that my estimated position was spot on. The second ship gave her name as *Fort St Joseph*.

The lull in the wind did not last long. It began to blow hard from the same quarter as before until it reached gale force. Unbelievably Corbeille ordered me to send more signals trying to bargain with both ships, playing off one against the other – not realising that they were in radio contact with each other. The outcome of all this was that since the *Fort St Joseph* was bound for Funchal she would wait for daylight and then take us in tow. *Godfrey B. Holt* was bound for England but said she would stand by till dawn in case assistance should be needed.

The two ships circled us for the remainder of the night. Wallowing broadside on in thirty-foot seas, some of them breaking over the well deck, *Woodbine* was in real danger of capsizing. At dawn *Godfrey B. Holt* wished us luck and departed to the north. In the course of sending all those signals in a full gale, at one juncture the side of the monkey's island against which I was leaning became the deckhead and I fell across the intervening space down upon the lockers on the starboard side and fractured a rib. It did not help future duties.

Beside me on the bridge Corbeille was muttering, 'I must think, I must think—' Obviously he was reluctant to accept a tow with consequent salvage claims outside territorial waters. Men were gathering at the foot of the ladder watching him with suspicion and mounting anger. The Skipper looked down at them and said, 'You'd *better* take that tow.'

At first light *Fort St Joseph* circled us closer and then came towards us from the north, upwind of us and to starboard. She signalled that it was too close for the Aldis (which worked by tilting a mirror) as the lamp was still visible at short range between the 'ons' and 'offs'. So I had to change to a simple flashlight torch, which fortunately had good batteries in it. There followed a signal saying a rocket would be fired across us, to which was attached a heaving line and towing hawser. Aboard *Woodbine* only eighteen men were fit to haul in a line. We had no winch. The first rocket fell short. The distance was about a hundred yards and at one moment the ship would be away above us and at the next we looked across her deck. She was in ballast and riding high out of the water with the tops of her propeller blades rising a foot or two out of the sea, causing a lot of splash.

She went round again and that time the rocket line fell right across our foredeck. It was gathered in rapidly by the Skipper and handed to the line of men ready to haul along the starboard side deck, after having been passed through the horns of the starboard fairlead. *Fort St Joseph* went slowly ahead, seamen passing out heaving line as we hauled in the rocket line. The small shackle joining the rocket line to the heaving line passed easily through the horns of the fairlead and then the job of heaving in the hawser began.

Yard by yard, foot by foot, the line came in. It was too heavy to lift out of the sea most of the time but occasionally, when the ships rolled apart, it did, and when that happened the Skipper jammed it in the fairlead with a wooden billet. In fact most of the hauling was accomplished by making use of this source of energy, heaving in on the fall and belaying on the rise. *Fort St Joseph* had to keep moving ahead to prevent the line fouling her propeller.

When the shackle joining the heaving line to the hawser came up to the fairlead it was too big to pass through. The Skipper shackled the hawser to the forward horn of the fairlead, fetched a chain, and shackled that to the after horn. He passed the rest of the chain across the deck and took a couple of turns of it round the mast.

Meanwhile the seamen of *Fort St Joseph* had fastened a stout hemp line to the bight of the hawser, some four fathoms outboard and hauled in on this to the port quarter. They made it fast so that the towline was held amidships over the stern in a bridle. We were thus being towed by the hawser from the middle of the bridle over Fort St Joseph's counter, made fast to one horn of our fairlead and then chain to the after horn – taken, by the Skipper's intuition, to the mast.

The tow commenced at about 10 a.m. The exhausted men had been hauling in, beating *Fort St Joseph's* forward progress, since early morning. By the time all was made fast she was towing us on about a quarter of a mile of hawser. *Fort St Joseph* signalled that she would go ahead at five knots and we set off to the south-west. Within half an hour the fairlead carried away, torn out of the edge of the deck by constant jerking, and became a link in the chain. With the hawser fastened to one side and chain to the other, it hung over the side and stayed like that until we reached port.

By this time the sun was shining and the seas were moderating. I went round the narrow deck to check on Joan and told her we were under tow. When she was conscious she could see out of the small porthole level with her face without lifting her head. I came out onto the deck, lay face down, and fell asleep.

After about an hour I was awakened by a jerk which was unlike *Woodbine's* movement in the sea. I looked up and saw that the hawser had either broken or come adrift as it passed out from *Fort St Joseph's* port quarter and was hanging down into the sea, held by the starboard line which had formed part of the bridle. Seamen were running aft to inspect. The remaining arrangement seemed to be standing up to the strain and thus we proceeded, at about four knots, hemp, wire, fairlead and the short chain to *Woodbine*'s mast, through the afternoon and all that night.

On the 27th of February we were in the lee of Madeira and at 11 a.m. we approached the Pontinha at Funchal. The tug *Gaviaio* came out to meet us and threw us a towing line. *Fort St Joseph* signalled us to drop our end of the hawser, which she hauled in, thus saving herself inport duty on it had it dropped into the harbour.

I sent a signal of thanks from Corbeille to *Fort St Joseph* and *Gaviaio* towed us to a quiet spot in the Pontinha where we dropped anchor. *Woodbine* was destined to spend several weeks at this spot, a rusty contrast to the sunlit breakwater and the harbour wall overhung with festoons of golden shower and the ascending rows of houses on the hillside of Funchal with green trees in the gardens and along the boulevard.

5. MADEIRA

JOAN:.

No tourist ever felt the intoxicating delight of our first look at Funchal. The blaze of bougainvillaeas, the white walls and orange tiles, the dark green background, and over all, the clear blue sky and glorious sunshine – it was like heaven to the scarecrows on the deck of the battered old trawler. We were a blot on the picture. We had tried to remove the grime in about a cupful of sooty water each. Fourteen days' growth of beard adorned the chins of the men. The women's hair hung in limp tails, streaked with soot like the men's beards. Peggy and Noëlle had changed into bright print skirts. I was too weak to stand unaided but I managed to do my hair after a fashion. It was long and had been uncombed for nearly two weeks. It took me an hour to brush and plait it.

All that time things had been happening. Captain Cuelho of *Gaviaio* came aboard with the port officials. He was a boisterously euphoric sea dog, with brilliant blue eyes and a voice suited to the bucko mate of an old Cape Horner. He had an impressive command of nautical English and he described how he had put out to look for us during the gale, searching all round the false position Corbeille had given, but having no radio, had been unable to find us.

The inevitable bumboats swarmed around, displaying gaily embroidered hats and bunches of bananas. The banana sellers did a roaring trade with the half-starved crew. Somebody had collected Sheila, washed and dressed her, and she was having the time of her life among the lads on the well deck, cramming bananas – at least one unpeeled – into her little tummy as fast as they gave them to her (there were no ill effects in spite of my frantic fears of typhoid and dysentery).

Robin had helped me to a chair outside my cabin. He was trying to shave in a bucket beside me.

'Darling, did you know your beard is red?'

'Yes, it would be,' he replied. 'When I tried to grow an operational moustache in Sharjah it was red – stuck out too. I had to give it up.'

'Perhaps the new baby will have red hair.'

I was thinking: What happens now? I *can't* go to sea in this ship again – I just can't. But I knew there was little hope of anything else, and I was too tired to worry any more.

Somebody shouted 'Doc!' and Robin, half his face lathered, disappeared round my cabin. He returned with a pleasant-looking man in civilian clothes. 'This is Mr Krohn,' he said. 'He is Reuters' representative here.'

Mr Krohn said, 'We've been waiting for *Woodbine*. We heard your husband's signals and we've been anxious about you. My wife has told me that if I found anybody on board who needed rest and care, I was to bring them home. Everybody agrees that you and the baby should be the ones to come. I hope you will.'

I tried to protest that three of us would be an invasion of his home, but he insisted. The kindness touched me so I was nearly in tears. I tried to thank him and accepted gratefully.

The press photographers wanted a picture of us on the narrow boat deck. Robin hastily completed his shave. Mrs Corbeille came to ask me if I was ready for the picture. She was dressed in a navy blue double-breasted jacket with brass buttons, a tight navy blue skirt, sheer stockings and high-heeled black suede shoes with peep toes. She wore a stiff-collared white shirt with narrow black stripes and a black tie.

'Do you like my outfit?' she asked. 'Captain Corbeille insists on it, you know. He says as the Captain's wife I must wear proper yachting costume. But I got a stripey shirt because otherwise I would look like a Wren.'

'You look very smart,' I said, straight-faced. 'I don't know how you manage it.' But I did know. She and Corbeille had a comfortable cabin, with wardrobe, washbasin, and a large mirror. Her hands were unspoiled by dishwashing in cold sea water, or peeling potatoes. I do not think there was any serious shortage of water in that cabin. However, I bore her no grudge. She had been nice to me and had washed and changed Sheila several times.

She helped me to the rail, where we stood looking down on the well deck. I had to be in the picture in case it got into the South African papers. My people had probably heard of the 'hospital case' and I did not want them to think it was me. I had not told them about the expected baby. Mother was anxious enough about us sailing in so small a vessel, though I referred to her as a 'converted trawler yacht'. However I knew I could not pull any wool over my father's eyes. As a seafaring man he knew exactly what we were undertaking.

Woodbine had been reported 'overdue' and then 'missing' at Lloyds. We knew they would be desperately worried.

Corbeille pressed to the middle of the group for his picture. Gripping the rail he swung his arm forward to display his gold rings and threw his head up like the Monarch of the Glen. His was the only recognisable figure in the published picture.

It was now several hours since we had come into the lee of the island and my sickness had stopped. I had eaten nothing and drunk barely enough to keep alive for fourteen days. I began to feel ravenous. I was afraid to eat too many bananas after such a long fast. I had two, but they did nothing to allay the pangs. There was nothing else to eat on board.

At last we were given permission to go ashore. Looking back at the little ship I wondered how we had survived.

'It's her broad beam,' Robin said, reading my thoughts. 'She's a drifter. Nothing else would have lived in that sea with no way on her.'

'I know,' I said, 'but just *look* at her!' Poor little *Woodbine* had looked bad enough in Southampton against the winter background. Under the clear sunshine of Madeira she looked appalling. Her grey paint had given way to all-encroaching rust. Great streaks and blotches of it covered her. I turned away. No ship should ever look like that, I thought, it's almost indecent to look.

As we drove through the streets of Funchal I found myself staring wolfishly at some children chewing buns and sweets and thought: How awful: you really could 'take candy from a kid'! Robin sent cables to South Africa and England and then Mr Krohn drove us to his house. Fairly high up on the hill we passed between heavy wooden gates in a high wall and entered the garden. It was beautiful – homely and exotic all at once. There were lawns and terraces, incense trees and jacarandas, paved paths and orchids. Orchids by the dozen – in the garden, in pots, in hanging baskets, growing on trees – orchids everywhere.

Mrs Krohn received us in the doorway and led us through a high, wide corridor into a charming pink drawing room. In my semi-washed condition I was installed on a brocade couch. An electric fire was arranged at my feet for I was shivering with weakness and hunger. Mrs Krohn said, 'You must have lunch first, and then you can have a warm bath and go to bed. I've sent for my son's old Nanny to look after your baby and you can have a long sleep.'

She left us alone in the delightful room. One wall was covered with a priceless collection of Royal Crown Derby china. There was a grand piano, and on a table nearby stood a bowl of pink camellias with the light behind them.

The next few days were a dream. Things were humming between

Corbeille and the owners of *Fort St Joseph*, their agents, and the British Consul. There had been a whip-round aboard *Woodbine* and a good sum was collected for the men who had handled the tow aboard *Fort St Joseph* to have a drink on us all. The Steward volunteered to take it aboard her. He took it all right, but not aboard *Fort St Joseph*. Only his adiposity saved him from retribution when the Woodbines found out. You can't beat up a mountainous blancmange.

The owners of *Fort St Joseph* were claiming a day's demurrage – £3000 – and Corbeille had no intention of paying it. I knew little of all this. I had been translated into Heaven. My six-foot-by-five cabin had changed into an enormous bedroom with a high ceiling and tall doors. In the afternoon shutters were closed over the windows for siesta and everything was peaceful and quiet. I slept, ate wonderful, carefully planned meals, and wallowed in glorious baths. With bath salts! and scented soap! There were cream and brown orchids on the dressing table too. Robin had bought them in the town because he had never had the chance to buy me orchids before. Sheila had been whisked off by the Portuguese Nanny and I had no responsibilities at all. At twenty-five one recovers quickly, especially in conditions like these. Soon I was able to get about again and we met some of the Krohns' friends.

Robin had been to the British Consul several times. Everybody was anxious to know what was going to happen next, and Robin was the natural spokesman. *Woodbine* would not be allowed to sail until Corbeille had paid the three thousand pounds to the agents. *Fort St Joseph* sailed the day after she had towed us in, her people unaware of our gratitude and the token of it pocketed by the Steward. Corbeille refused to pay, and there we all were – waiting for something to happen.

What happened was the arrival of H.M.S. *Duke of York*, showing the flag. The Admiral sent for Robin and asked him about *Woodbine*, Corbeille, and the situation. He did not see Corbeille. 'Now isn't that a strange thing,' Noëlle said, laughing. 'Have you noticed that there is no gold shining on him now?' We had. For the duration of the battleship's visit the Master of our ship wore no gold braid. It was at this time too that we heard about his ex-partner now expiating his sins at government expense. Wherever any of the Woodbines met rumours and scandals flew fast. Some of us were still wondering about our sudden hurried departure from Southampton . . .

With the Krohns Robin and I were invited to the ball at Government House. Robin crawled into the dark confines of *Woodbine*'s hold to hunt among our boxes for evening clothes. Peggy and Noëlle were on deck when he popped up in his old uniform jacket and pointed to the three rings on the cuffs. 'These are real,' he said, and popped down again, leaving the girls laughing. It was not pleasant hauling boxes about to find the right one, but he emerged eventually with his tails and a dress of mine. It was only when he came back to the house that he told me about his broken rib.

Robin looked fine in his pre-war tails but I, though I had always been fairly slim, was now too thin for my dress. Mrs Krohn sent for a sewing woman who took it in. Many months later when I thought to wear it again I could not believe how thin I had been at Madeira. It was hardly surprising. What did seem unreal was to be going to a ball at all, in evening dress, with orchids in my hair and Mrs Krohn's perfume behind my ears, so recently filthy with coal dust and neglect.

Government House – the Governor's Palace – was palatial even by standards more critical than we had in *Woodbine*. Perhaps it was the sharp contrast that made it so memorable. It had been a long time since we had been to a grand occasion. I thought of all those cramped lodgings and the tokolosh in the scullery pump – now all that was unreal too. The ballroom was beautiful with good proportions and silk-covered walls. We stood in a group with the Krohns' friends, talking and watching the crowd. I was still a bit shaky but loving all the colour and movement. Then the Naval officers arrived, together as they had come ashore. As usual, the plain black, white and gold of Naval mess dress made other uniforms present look ornate. We looked at the group in the doorway with sheer tribalistic pride. Surely they could not *all* have been tall and fair and handsome? Of course not – it was just the magic of that unbelievable evening. After that moment the evening passed in a whirl – not much dancing for me but a great deal of talking and laughter. They asked about *Woodbine*. We said she was a good little ship, and they must not, please, judge her by her condition. They had heard about Corbeille. That accounted for some of the laughter. Sailing under royals made of army blankets! And the uniform! Our trawler and her motley crew had obviously been discussed in their wardroom.

They told us that our gale had indeed been a bad one. 'Knocked *Vanguard* about, with the Royal family aboard—' 'I believe they had a piano adrift—'

The Royal family were on their way to South Africa at the same time as we were running out of coal off Porto Santo. The Naval officers agreed that *Woodbine* had survived because she was built as a drifter.

All the time, when there was a momentary break in the talk and music, the thought would swim to the surface of my mind – What do we do after all this? And I would put it down again and try to forget our problems.

It is all so long ago now. Problems are resolved and you find yourself with a new lot, on and on to the end, when the next generation starts all over again.

Nothing had been resolved when we attended H.M.S. *Duke of York*'s cocktail party. This was the era of the 'little black dress'. I had had one once, but Robin loathed it and I gave it away. Cocktail dresses were usually of marocain or some smart material never worn by day. I could not let Robin go into the hold again with his broken rib to rummage for a suitable dress. I wore a pale beige linen, pinned orchids on the shoulder, and got several compliments on it. Well, it did rather stand out among the little blacks and pearls.

I went out to the ship in her boat with some trepidation but I was not sick. Several other people looked a bit green, so I was slightly encouraged. Standing on the vast quarter deck of the battleship I thought of poor little *Woodbine* again. She had once worn the grey paint that was now so dilapidated, as a humble unit of the Grey Funnel Line – when those enormous guns behind us were blasting out in deadly earnest – when I had been nursing survivors of H.M.S. *Kelly* and so many others, and playing tennis with young petty officers of H.M.S. *Barham* before she went to her terrible end. I wondered what Corbeille was doing all those years – he never mentioned war service – and I hoped *Woodbine* would get safely to South Africa and be bought by honest fishermen there.

* * *

Mr Krohn was not only Reuters' agent. He owned a factory where the exquisite embroidery of Madeira was designed and made. He took us and the girls around it one day with some of the officers of the *Duke of York*, and we were astonished at the way it worked. On the top floor were the office and design studio full of reference books, files of motifs and ideas to inspire new designs. I could have spent hours in there. On the lower floors we saw the progress of the work from the transferring of the designs to the finishing. We saw everything except the actual embroidering, for the material with the designs was farmed out to peasants in the mountains. Men, women, girls and youths all worked on the embroidery. A large piece would have several members of a family working on different parts at the same time. Mr Krohn told us of one enormous tablecloth that had been worked in this way, but unfortunately when the elaborate design that started from both ends met in the middle it was found that one half had been embroidered on the wrong side. They had unpicked the half and re-done it.

We saw completed embroidery being delivered by peasants. They handed in rolls and bundles of what looked like filthy rags, took their money, and went out with fresh material. Mr Krohn laughed at our expressions when we saw those bundles. They were so crumpled and dirty it was impossible to connect them with the famous Madeira embroidery we knew. The next place he showed us was the laundry, where hefty women were washing and ironing returned embroideries by hand, with what seemed to us a lot of vigour for such delicate work.

The next stage astonished us even more. Women were working on the now immaculate cloths, cutting out parts of the design to make the traditional open-work patterns. What amazed us were the scissors and the speed at which they worked. Great big blades that a tailor might have used flashed and chomped in the strong peasant hands – and produced the final dainty and lace-like patterns. The girls and I agreed that we would timidly have used the smallest scissors we could find and have taken hours to do the job they did in minutes.

After the final checking and pressing the work was taken to the factory showroom and there we saw work that had been commissioned by famous people or institutions from all over the world. These were not the luncheon sets and tablecloths that are offered to tourists. We saw one tablecloth about fifteen feet long, of white organdie, embroidered on the underside with 'shadow-work' that gleamed through the thin cloth. The design was hydrangeas in natural colours and size. The pinks, blues and delicate greens shimmered along the sides in a border about four feet wide. The Hollywood hostess for whom it had been created had had her china made to match it in the same design. After seeing this and other wonderful things we bought table mats for ourselves before we left. The design and workmanship were just as good as in the fabulous things we had admired in the showroom. We have ours still – wheat ears and flowers on pale yellow organdie, veterans of many dinner parties over the years.

Mr Krohn's family had not always been in the embroidery trade.

Before the revolution they had supplied Madeira wine to the Tsars of all the Russias. One of them had been so pleased with it that he had sent Mr Krohn's grandfather a beautiful gold snuff-box with an enamelled landscape on the lid. When you moved it in the light, rays fanned out over the mountains as though the sun were rising. The house was full of fascinating relics of the wine trade and Mr Krohn told us many anecdotes of earlier times.

6. IMPASSE

When the naval visit was over Robin and I and Sheila moved back aboard *Woodbine*. It had been wonderful but we could not accept hospitality indefinitely now that I was better. Conditions on board were as bad as ever. There was little food, hygiene was elementary and morale low. Though the ship was not under arrest the authorities had removed a crosshead of the engine. Corbeille had been to Madeira in another vessel the previous year and had left bad debts in the town so they were firmly on the side of the *Fort St Joseph*'s agents in their claim. Corbeille and his wife were living ashore somewhere but the crew remained on board, at least to sleep. Britain only allowed her citizens to take £75 each beyond her shores and nobody knew how long the impasse would last.

There was no hope of ships going to South Africa. They were full of emigrants and their crew strength was complete. Some of our professional seamen signed off at the Consulate and signed on in ships going to Canada and other destinations. The amateurs remained, with, fortunately for all concerned later on, the stokers.

All *Woodbine*'s remaining people were now back on board except for her Captain and the Captain's wife. The ship's stores were being depleted but the days passed and still nobody knew when or if we would ever resume our voyage. Ships came in on their way to Cape Town and we watched them sail out again with resignation. We knew the Consul and the shipping agent were enquiring for vacant berths in every one. One day I was sitting outside my cabin writing to my family and watching one of the Union Castle mail ships leaving, when Peggy and Noëlle came aboard. They had been to the agent's office. 'They're very annoyed,' they reported. 'That bloody ship has sailed with dozens of empty berths!' Apparently weather in England had delayed trains that were bringing passengers and they had missed the ship. The Madeira people had only found out when she was nearly due and there was little time to arrange for the stranded people from *Woodbine* to take some of the empty places. The Union Castle refused to take any of us although we were not asking for free passages. We watched her disappear with hard feelings indeed.

Among the mixed bunch on board was one man who never grumbled. He was an old retired army doctor who kept entirely to himself. He took no part in the endless speculation and left the leadership of the rest of us to Robin. He had all his belongings in the hold, including ancestral portraits by famous artists, worth a great deal of money. (No wonder Corbeille changed his mind about burning our baggage that night off Porto Santo!) The old man looked rather frail and must have been desperate to get to South Africa to sail in *Woodbine*. Except about his things in the hold he never confided in Robin or the Consul.

One morning, to everyone's surprise, sacks of coal arrived in a barge and were being loaded. Corbeille and his wife were back on board and officers were bustling about. The Chief appeared and looked around with a conspiratorial grin. 'Foxed 'em,' he said triumphantly. 'Me, I had a spare crosshead all the time. Only wanted the coal.'

'Has Corbeille coughed up?'

'Not that I know of,' he said. 'The old Doc paid for the coal.'

'Well, we can't sail till that claim is settled.'

'Told you I had a spare crosshead. Only got to get outside the three-mile limit and they can't touch us.' He went back to his engine to start raising steam.

It was ridiculous of course. A ship carries her debts with her – and anyway the authorities could not miss all the activity and smoke beginning to appear from the funnel as the stokers raised steam for the old reciprocating engine. We watched with interest, certain that *Woodbine* was not going to sail.

Sure enough, things were happening ashore. A boat came out, full of armed policemen and rowed around the ship several times, looking at her closely. Apparently satisfied that we were up to no good they came alongside and clambered aboard. One of them clutched a paper and they all had rifles. Coaling ceased forthwith.

'They're going to arrest the ship,' Robin said. 'That must be the warrant to nail to the mast.'

'They can't nail it,' I said. 'It's a steel mast.'

'Well, they'll have to tie it on with string.'

That is exactly what they did, and though less dramatic than the old way, *Woodbine* was effectively under arrest.

'Do you think those guns are loaded?' I asked Robin as grinning gendarmes swarmed over the ship. One fat one was beaming at Sheila.

'They're sure to be,' Robin replied. 'All this *is* serious, you know. I just hope Corbeille isn't going to do anything stupid.'

'Well, they're jolly casual with them if they are loaded,' I said, picking up Sheila. 'I'd have got a rocket from my uncles if I'd carried a gun like that, even unloaded. Let's get below until it simmers down.'

We took Sheila down to the mess and the others soon followed. Some of the policemen remained on board, taking watches to guard poor little *Woodbine*. Sheila adored them. Soon the jolly fat one was walking up and down with Sheila on one shoulder and his rifle slung over the other. He told us in halting English that he had grandchildren himself.

So our strange life went on. There was little respect left for the Master of the Ship. Peggy, Noëlle, Robin and I were sitting on the strip of deck outside my cabin one evening, watching the lights on the harbour. Sheila was asleep in her cot behind the barrier in the cabin doorway. There was a lot of noise on board and muffled shouts from below. Then some of the boys came clattering up the ladder.

'Come on – we're all on the engine room hatch—'

'It's a party. We've busted into Corbeille's stores!'

'You mean the bonded stores? All that hooch?'

'Sure. It's not going ashore. The party's here on board.'

'He owes it to us. He was only going to flog it.'

'That's right – he's a bloody spiv.'

The four of us joined the party with glee. It was in full swing. Bottles of spirits were being passed round – whisky, gin, brandy and liqueurs. It was at the jolly stage – everybody full of guilty delight at getting even with Corbeille. Some of the boys had got into a boat and were rowing round the ship, their oars splashing in and out of light sparkles on the water. English songs roared out across the harbour. Forced to be rather abstemious, I sipped my well-diluted drink and watched the scene with an artist's eye – but one would need the skill of Rembrandt to catch it. Robin sat with his arm round me and I thought how lucky I was to have a husband who would never land us in suburbia.

The party was getting uproarious when I judged the right moment to leave the boys to it. 'Must go and see Sheila,' I said, getting up, and Peggy and Noëlle said they would go and see her too. After a while Robin joined us on the boat deck.

'What a party! There'll be the most colossal thick heads in the morning.'

'Corbeille will be absolutely furious,' Noëlle said with satisfaction. 'I wonder what he'll do.'

Gradually the party flickered out, leaving only the gasping puffs of the un-pumped pressure lamp that still flared from behind the funnel to break the silence of the harbour.

If Corbeille was furious he had sense enough not to vent it on his hostile and hung-over crew. We heard no more about the party, or the cases of spirits left over from it which had vanished into the fo'c'sle, never to be seen again.

After a few days Robin declared that the lack of hygiene was so bad that we and the girls would have to move ashore. He wrote an official letter to Corbeille saying that he would not be responsible for the health of the crew under the conditions on board. Corbeille's dislike of Robin was turning into hatred by now.

ROBIN:

We had not yet found a place ashore when the Skipper came to me, furious. 'Doc,' he said, 'Corbeille is going to take paying deck passengers down the African coast to Lobito – recruits for the mines. He has ordered me to arrange for the construction of latrines on the deck!'

I was appalled at the prospect of a large number of men, cooking, eating, and sleeping on the decks of *Woodbine*, under ghastly conditions for both them and the crew, in all weathers. Then I said, 'If he does this he'll lose the membership of his yacht club. They'll be interested to hear about it.'

I composed a cable to the Commodore of the yacht club saying: 'Your member Corbeille intends taking paying deck passengers down the West African Coast.' I signed it with my name and profession. As I had expected, the yacht club reacted quickly.

I had rowed ashore and was walking around the deserted end of the Pontinha, beneath the high walls and near the brink of the quayside, when Corbeille came up briskly behind me and snarled, 'What do you mean by telling my yacht club that I intend to take deck passengers?'

'Well, it's true, isn't it?' I asked.

'It's no business of yours what I choose to do. I am the Master of the ship. I have a bloody good mind to teach you a lesson.' He clenched his fists threateningly.

'O.K., come on,' I said, squaring up to him. He turned on his heel and strode away toward the town and I was left regretting a lost opportunity of dumping him into the harbour.

JOAN:

This incident lost Corbeille even more respect from his crew. I heard one comment: 'He's all bluff. Doc's muscles don't come off with his jacket.'

Robin found a room for us in one of the big hotels. As it was so soon after the war there were no tourists and the hotels were almost empty. The management let us have a small but comfortable room for an almost nominal fee. The whole staff took an interest in us and the chambermaids adopted Sheila almost completely. We would see her being carried off to the kitchen by plump jolly women who crooned over her fair curls and called her 'Sheilita'. Sheila of course adored them and one day we found her scooting down the corridor at the rate of knots on her little pink potty looking for her friends.

Robin had interviews with the Consul and agents and got to know many of the officials. He was the undisputed leader of *Woodbine*'s people, and the younger ones came to our room to tell us all the news, rumours and scandals – sometimes late at night. The boys patronised a rather nice night club. We went there ourselves sometimes. Robin used to go with his three young women and our party caused a lot of interest wherever we went. Noëlle was blonde and blue-eyed – she must be the mother of the beautiful blonde baby – and yet the dark brunette was obviously the official wife, and she was friendly with the blonde mother of the child – and was the other woman yet another wife? Well, everybody knows the English are crazy – but surely the young doctor's menage was quite something, even for them. We got smiles and greetings and puzzled looks on all sides.

Our night club was respectable. Most of the girls had chaperones and the dancing was almost demure. For us it made a pleasant and not expensive evening. We had no shortage of baby-sitters for Sheila. I did sometimes wonder whether she spent her evenings in the kitchen and how much Portuguese food she stowed away in our absence.

One night we had gone to bed early when the boys came thundering on our door, bursting with excitement. Somebody had fetched Peggy and Noëlle from their room on the floor below and they too were agog.

'Corbeille's inside! He's been arrested!'

'Good Lord!' Robin and I sat up, very wide awake. 'What's he done?'

'Tried to clobber a Portuguese officer in the night club!'

'We saw it—!'

'Threw a glass of whisky in his eyes—'

'You should have seen it!'

'Tell us properly then,' said Robin. 'Why did he attack the officer in the first place?'

'Didn't like the way he was dancing with his missus. She'd had several dances with him and Corbeille got jealous.'

'Were you all there?'

'Oh yes! It was terrific. Everybody was standing up and suddenly each of us had a Portuguese bloke standing beside him with a knife in his hand.'

'They just said quietly "You don't want trouble for this man?"'

'And *we* said: "Go ahead chums. He's all yours. You don't have the death penalty here and we do."'

'And then we all walked out—'

'Did you see Corbeille arrested?'

'No. When we left there was a big bloke holding one of those marble-topped tables over his head—'

'But the gendarmes were coming. We buggered off and came along to tell you. What happens to us if he's put inside properly?'

What indeed?

Next morning Robin went to the Consul once more. That patient man was doing his best to soothe things down. Corbeille would probably be fined for assault and breaking the peace. The Portuguese were being very decent about it. Funchal knew Corbeille and only wanted to be rid of him but were sorry for his luckless crew and would be as lenient as they could.

All was quiet for a few days. Then Robin got an agitated call to the ship. 'He's tried to kill her!'

'What?'

'He was strangling her – she's hurt, Doc. Her neck's black and blue. You'd better come.'

The poor little Captain's wife had been beaten up and the marks on her neck showed she had been very nearly strangled. The Master of the Ship, foiled in time by the Chief – luckily for himself as well as her – had gone blazing ashore and nobody knew where he was. It was no longer funny. He had knocked his wife down and kicked her savagely. Robin was horrified at the bruises on her breasts and neck. 'She wants to stay in her cabin,' he told us when he came back. 'She feels safer on board with the crew and they can see him coming by boat. I talked to her like a Dutch uncle and told her to leave him. It's not safe. With my report on her condition she'll get a divorce.'

It was a shaken group that gathered on the hotel porch. The Captain's wife was not a joke any more. They were sorry for her now. Still Corbeille had not reappeared. He did though, the next day. His wife was on deck, wearing a high collar and long sleeves. Corbeille came aboard and locked her in the cabin. The men watched his every move.

'She's under arrest,' he informed them. 'Nobody is to speak to her or take her food or drink. *Nobody!*' He glared round, and then got back into the boat. There was silence as he went ashore and then the men went up to the boat deck and knocked on her cabin door.

'Mrs Corbeille – are you all right?'

'Yes, thank you. You're not supposed to speak to me.'

'The hell with that – he's gone. We're going to let you out.'

We heard all about it later from our young reporters. 'You know,' they said, 'she's all right. She's got guts. She wouldn't let us disobey Corbeille. She had a spare key – that made us laugh a bit – and she let herself out.'

'We had a dinghy alongside and we were all set to row her ashore but she said she wasn't going to get us into trouble. She rowed herself ashore – with all those bruises.'

'We told her to go to the Consul or come here to you.'

The Captain's wife was a heroine now. She had won the hearts of that mixed lot of men and Corbeille had better leave her alone.

'She hasn't come here,' Robin said, and went to the Consulate again. She was not there either. The silly girl had gone to the one woman on the island who was, for some reason, ostracised by the British community. This woman persuaded her not to go to the Consul. She should forgive and not try to divorce her husband (for attempted murder?) We were all relieved when we heard that she had flown to Lisbon en route for England.

The boys had another piece of information for us when next they came to our room.

'You know, we forgot to tell you in all that excitement that before she left the cabin Mrs Corbeille broke into his locked drawers and got all his papers. Some of us saw them.'

'Do you know what he got for that minesweeper he took to the Cape last year? £35,000!'

'So what's he going on about £3000 for?'

'He only paid £2000 for *Woodbine*, so he got a good profit.'

'Don't forget all the fares we paid . . .'

'And the other bods last year . . .'

'Well,' said Robin, 'if he isn't going to fork out we can't make him. But it can't go on like this.'

'That bastard! – if we get to sea again in *Woodbine* he'll go over the side one night.'

There was a silence. It was not funny. Corbeille had brought latent violence on board.

'It wouldn't be worth it,' I said.

Silence. Then:

'It would be dead easy.'

'He's a bastard.'

'Remember how he beat up that bloke in Southampton?'

'And he was too bloody mean to take a tow and nearly killed us all.'

'And his own wife!'

'He's nuts,' I said, 'and definitely not worth the risk of murdering. He's the murdering bastard, not us.'

Peggy and Noëlle backed me up and then Sheila came in with her chambermaid and everyone laughed.

When we were in bed that night Robin said suddenly:

'It would be dead easy.'

'Robin, you're the only man with a profession amongst those men. You'd be the one open to blackmail.'

'I'm only saying it would be easy. Darkness, low rails . . . Seriously, he's pushing his luck, you know. One night somebody really may give him a shove.'

'O.K. – so long as it isn't you. How the hell did we get onto such a subject?'

In the morning dark thoughts vanished in the sunshine – but I was afraid a nasty seed had been sown and tried to suppress another niggling worry.

After the departure of his wife we saw little of Corbeille. He was still living ashore but sometimes appeared on board and shut himself in his cabin. He ignored everybody and was brusque and secretive when asked how things were going. It had become very clear to his crew that the Master of the Ship really was a plain spiv.

The next incident came out of some trouble he got into with the Customs. Our information service soon arrived, agog with the latest drama.

'He's done it *again*! He's hit a Customs officer!'

'Hit him in the face – an oldish chap too—'

'We heard it was something to do with a gun.'

'My God, has this lunatic got a gun?'

'Shouldn't think he has now—'

'They sent for the Consul of course—'

'The Consul is in bed with flu,' Robin said. 'He's been laid up for days.'

'Well, he got up and went to the gaol, or charge office, or wherever the gendarmes took Corbeille.'

'It's too much,' Robin said angrily. 'The poor man is ill. He's been extremely decent all along. I just hope he doesn't get pneumonia.'

This time it took longer but the Consul got Corbeille out again. We saw even less of him after that. Rumour and speculation continued. One day our news hawks arrived bubbling over.

'You know why we sailed so suddenly from Southampton?'

'We've often wondered.'

'Well, the cops were after Corbeille. He got tipped off that time he dashed off to London—'

'And came back in such a hurry he wouldn't wait for steam coal—'

'Well, one of the lads has got a letter from a pal. He's made enquiries about Corbeille and found the police were making them too. It's in the letter.'

'What for?' Robin interrupted. 'What were they after him for?'

'Bigamy.'

Nothing could surprise us after that.

* * *

Nearly six weeks had passed since *Fort St Joseph* had towed the battered trawler into Funchal. The Consul's flu was worse after he got up to help Corbeille. 'He should have left the bastard inside until he was better,' the boys said. We all agreed that Corbeille did not deserve such service.

As Robin had said, this sort of thing could not go on. We were all running out of money and there seemed to be complete stalemate over the salvage claim. Corbeille obviously had no intention of paying and nobody could guess what was going on in his tortuous mind. As soon as he was about again the Consul spoke to the Anglican vicar and discussed the situation.

'These people are in a bad plight through no fault of their own.' Although they had signed articles as crew they had actually put a lot of their savings into the voyage. They could get no more money out from Britain and would soon be unable to buy food. He could of course send them back as distressed British seamen, but he would only do that as a last resort, he explained.

Like everyone else in Funchal the vicar knew most of the story and was sympathetic. Perhaps he could intervene on compassionate grounds? He and Robin met in the Consul's office and worded a cable to the owners of *Fort St Joseph*. Robin paid for it gladly. In it, the vicar asked them if, in view of the plight of the passengers and crew, they would accept a token payment; and release the ship.

The reply came on the following day. We were astonished at the generosity of the owners, whose claim had been moderate to begin with. They pointed out that although they considered the cable to be inaccurate and unnecessary, they would accept £300 instead of the £3000. The only person who was not grateful was Corbeille. He never even thanked Robin for paying for the cable. When the £300 had been paid – Corbeille could not refuse that once his hard-eyed crew had heard about it – the policemen gathered up their rifles and rowed ashore, waving friendly farewells.

Good steam coal began to arrive from the agents. As the bags came aboard Corbeille stood by with a portable weighing machine, pulling out every other bag and making sure that the old-established and highly reputable firm of Blandy Brothers was not cheating. 'Judging everybody else by his own lousy standards,' one of the boys said, watching the ludicrous figure with its gold rings and peaked yachting cap. Nobody helped him weigh the bags.

Coal and water had been taken aboard, and presumably adequate provisions. Parts of Woodbine had been painted and she looked better than she had on arrival, but it was a pity that the paint had been in cheap job lots, because no two had been quite the same colour. There were shades of blue contrasting with the original grey, but it had covered some of the rust. She was a game little ship, but I dreaded sailing again with Corbeille.

While all this was going on we were still staying in the hotel. The very few guests in there had been following the *Woodbine* story with interest. Inevitably I had been alone quite a lot as I got tired quickly when the others went to town. A friendly Yorkshire woman was making beautiful tatted lace as a hobby and she taught me how to tat. I used to sit in the billiard room sometimes, tatting yards of lace edging while Robin and the girls played snooker. The barman used to make me fresh orange drinks and when I thanked him he reminded me that Portugal was England's oldest ally.

We got the local version of the incident in the night club from him. They had all been amused at the way our people politely left their Captain with an outraged Portuguese holding the table over his head, saying there was no death penalty in Madeira. When they heard that we were going to sail again our friends in the hotel were very concerned, especially for Sheila. The maids looked at us as if we were condemned to death.

ROBIN:

Remembering the incident of the American ship steaming unconcernedly past while we were drifting away from Madeira without coal, I was determined that we should have some really effective rockets on board.

After making enquiries I located a shop which dealt in articles such as fancy dress, theatrical make-up, masks and other paraphernalia connected with merry-making and carnival. They also made fireworks to order. I went in and consulted the manager whose English was a little better than my Portuguese. Oh yes! They could supply *'fogos de artificio'* – the ones that mounted – psst – bang! And did I wish for *detonantes* or *illuminantes* – and they would supply them complete with sticks. They were astonishingly cheap. As far as I remember they were about four for a pound. I ordered three *detonantes* and three *illuminantes*, the first to attract attention of somnolent (or absent) lookouts, and the second to establish our location.

Two days later I called to collect my purchases. They were ready – all six of them! The business ends were about a foot long and the last three inches consisted of a tough blue paper bag tied to the top end of each sturdy two-metre stick. The bottom end of each *fogo* consisted of a twist of paper, out of which protruded what looked like an inch of impregnated string.

I rowed them aboard *Woodbine* in darkness and stowed them away with the connivance of the Skipper. Corbeille never knew we had them.

7. LAS PALMAS

JOAN:

*W*oodbine was cleared for Las Palmas. We had said goodbye to everybody and tried to thank them, especially the Krohns, for the kindness they had lavished on us. Quite a lot of people came to see us off. A woman from the hotel took my hand and said, 'I think you are very brave.' Full of apprehension, I said, 'No – just desperate.' I was terrified of being seasick again.

Many of the crew had fortified themselves with Madeira wine and were lugging wicker-covered bottles aboard for future feasts. A quiver of the deck showed that the Chief had got steam up and that we were ready to go. Robin was not yet aboard.

Woodbine's siren would have done justice to a great liner as Corbeille paced the bridge with a face like thunder. He would dearly have liked to leave Robin behind but he was the radio operator, and besides, the crew would not have allowed it. The siren stopped as the agent's boat was seen tearing towards us. Robin climbed aboard clutching Briefcase, waved to the bridge and came into my cabin.

'I'd lost it again,' he said shamefacedly. 'Thank Heaven the clerks found it in the agent's office.'

Words failed me. 'Only one item of importance and easy to watch.' Obviously it was easy to lose.

'Let's not make a habit of it,' I said, holding him tight.

With much hissing of steam the anchor rumbled in. Water churned under the counter and after six weeks of unnecessary delay, once more a free ship, *Woodbine* slowly gathered way. Her siren roared again, this time in farewell to the people on the quay. I was sure that Corbeille had not ordered *that*.

The shining steel crosshead which had been commandeered by the Port Authority had been returned on the day the fresh supply of coal had been shipped, and it (or the spare that the Chief had chuckled about) was in place. But somehow a gasket somewhere in the arrangement lacked steamtightness, and at every revolution hissed loudly and some mechanical joint, insufficiently tightened, clanked. The resultant sound became a password between us all – 'Pss-kerjoink, pss-kerjoink!'

Woodbine hissed and clanked her way to Las Palmas.

I watched Madeira fade in the distance, saying to myself: 'I will not be sick. I will *not*—' *Woodbine* was rolling over the Atlantic swells again, but the sun was shining. 'I will *not* be sick—' Before Madeira was out of sight the whole miserable business had begun again. I lay on my bunk with a tin can and got worse and worse.

The passage to Las Palmas was only three days, and the weather remained good. Normally I would have enjoyed it but obviously pregnancy had totally upset my sea legs. Even three days is a long time to vomit non-stop – with what the old sailors called 'the dry 'eaves'. This time I was even worse than before for when we reached Las Palmas it did not stop. For hours after we arrived it went on. Once I woke up to see the resplendent figure of Corbeille staring at me from my doorway. *Woodbine* was in port, perfectly steady, and I was still being sick. He said gruffly, 'Do you want anything?'

'Hot water. To drink.' I had a craving for it. Corbeille brought it to me himself. Peggy and Noëlle could not believe me when I told them later. 'He must have thought you were dying and didn't want a corpse on his hands. You certainly looked like it.' It was probably the most surprising incident of the voyage.

He was soon back in character. 'Nobody is to go ashore,' he announced. 'Everybody is to remain on board.' Then he climbed into the harbour launch and went off in full fig. Of course as soon as he had disappeared the men whistled up the watching bumboats.

Robin and the girls were very worried about me. I was still being sick, and looked, they told me later, absolutely ghastly. 'She's got to go into hospital,' Robin told them, 'This is more than plain seasickness.' I heard them talking, but I was far away and feeling dreadful.

I felt the cool air as they half-carried me on deck and down the ladder, and into a boat. I heard Robin call the boat and knew it was night. Then Robin held me while the boatman rowed ashore. I was delirious – confused and frightened. Suddenly I saw another boat close by, full of British sailors. Jerking up, I babbled, 'Oh look – the Navy's here – just like bloody Narvik – oh thank God – it's all right now—'

The girls and Robin held me firmly and quietened me down. The boat was from a Spanish Naval vessel and there were four sailors in it with coloured ribbons on their shoulders – a 'paying off' custom, we were told later.

Somehow they got me ashore. There were no taxis and I was unable to walk. Robin dashed off to find one and a hospital and I slumped on the pavement. Peggy and Noëlle sat beside me to keep me from falling into the gutter. 'There we sat in the dark' they told me later, 'looking like a couple of tarts propping up a drunken pal. Fortunately Robin came back before there were any complications.'

I came to about twenty-four hours later with a pretty red-haired nursing sister looking at me. I was in the British Seamen's Hospital and Robin had found a cheap pension for himself and Sheila near the docks. I was feeling better but terribly weak. The new baby was still with me.

The British nurses were very interested in our story, especially when we told them how kind the British residents had been to us in Madeira.

'Nobody gets any help from *our* British residents,' they said rather bitterly. 'We don't even know any. We are "working girls" and beneath notice. We have no social life at all. Even in the war the nurses had no friendliness from our compatriots here, and some of the girls hadn't been home since it started.' It was incredible after Madeira. They were such nice girls, young and attractive.

'That's why, of course,' Robin said.

ROBIN:

That night, after Joan's admission to hospital, I returned on board and found Mrs Hoppy looking after Sheila. The crew too had returned by various means as they had nowhere to sleep ashore. It was unanimously agreed that we should all visit the Consul the following day to ask to be discharged. I was in favour of this. There had been a night on the passage to Las Palmas when I came out of the starboard door opening onto the small after deck – now somewhat larger since the decking over the tiller mechanism had gone up the funnel. I stopped short. Within a yard of me in the darkness Corbeille was leaning over the side with his hand on the metre-high bulwark, looking forward towards the bow wave. I retreated into the doorway. It would have been the simplest thing to push him overboard. He never knew I was there. It could have been somebody with fewer inhibitions than I had.

Next morning I went ashore with them in some of the bumboats who took an interest in our goings-on. My arrival at the Consul's office, which was on the southern outskirts of the port, was delayed by my telephoning the hospital for news of Joan. When I arrived at the office I came into a waiting room, one door of which opened into the office proper. The room was crowded with the rest of the crew.

Just as I arrived the Consul's door opened and he invited us in. I forget who acted as spokesman. He put the case clearly. The crew wished to be discharged. It was unanimously agreed that Corbeille was incompetent and unfit to command a ship. He had put our lives in danger and would do so again. They would sail no further with him.

The Consul replied that there was no problem. Those of us who had paid Corbeille were free and he would officially discharge the paid hands. While he was explaining this the door burst open and in strode Corbeille in glittering gold-ringed monkey jacket and yachting cap.

'This is mutiny!' he shouted. 'I demand that you arrest all these people and have them put in gaol. I am the Master of the ship and they are under my orders.'

The Consul interrupted him quietly. 'Mr Corbeille, this is not your ship. It is my office. This file contains a full report from my colleague in Funchal. These people are free to leave your ship and I am about to record the fact officially. You have no authority whatever in my office.'

As we filed out we heard him say, 'Stay behind, please, Mr Corbeille.' We never knew what transpired in that interview but shortly afterwards Corbeille left for England in a Dutch ship. He said he would return after making financial arrangements – but the crew took that with the proverbial pinch of salt.

Joan and I never saw Corbeille again.

JOAN:

We could only afford the hospital fees for three days, and when I came out Sheila had enteritis. The room Robin had found was very small, with a blue tiled floor, a narrow iron bed with a coir mattress, a chest of drawers and a large picture of a blonde woman rolling her eyes to heaven, with flames coming out of her exposed heart, her expression bewildered but patient. He had made up a bed for Sheila in one of the drawers from the chest. She was lying listlessly in it, her little face pinched and pale. She had violent diarrhoea but was not vomiting and was taking boiled water. 'She's not dehydrated,' Robin reassured me.

He had to go to the Consul's office that afternoon, and I sat on the hard bed watching her, horrified at the rapid change, from a pink-cheeked active child to the wan little figure whose big eyes looked at me dully. After a while I changed her nappy. It was full of blood. At that awful moment there was a knock at the door. It was the young Third Engineer looking for Robin. He looked at my stricken face.

'What's the matter?'

'Get Robin *quickly*. I think Sheila's dying. He's with the Consul. Please *run*!'

He ran all the way and burst into the office. 'Doc – quickly – Sheila's in a bad way and Joan is frantic.'

Robin rushed back. 'Has she vomited?'

'No.'

'Right. As long as she takes water she'll be all right. I'm going to find sulphaguanadine but I won't be long.'

Sheila's eyes had gone glassy. She lay still in the drawer and it looked to me like a coffin. It seemed a long time before Robin came back, having managed to make himself understood to a chemist who spoke no English. He crushed a tablet in boiled water and got her to swallow. Then he sat beside me and we watched her.

An hour later Robin said, 'She's kept it down. I think it's all right.' He put his head in his hands. 'We've got to get out of this mess soon.'

The Consul was trying his best but it was the same old story. Twenty-four hours after Sheila had frightened me so badly she was smiling and taking her dried milk feeds. In three days she was toddling about as usual though still pale and thin.

'I can't believe it,' I said. 'She was absolutely unconscious and look at her now.'

'Children are amazing,' Robin said. 'They go down so appallingly quickly, but they pick up again just as fast. It's because she never vomited that she's doing so well. That's why I tried the 'guanidine first, before the awful business of drips.'

Sheila had a Spanish fan club as enthusiastic as her Portuguese one in Madeira. The difference was that we were no longer in a large, nearly empty hotel, but living in a dockside pension for the equivalent of two shillings – ten new pence – per day. The room was clean but opposite its door was the communal lavatory for the floor and it made its presence felt. Goats and chickens lived on the flat roof, under the washing on the clothes lines. Down below was the dining room where one could get a nourishing meal of dried cod.

Peggy and Noëlle were still living in the Black Hole on board, with Mrs Hoppy. We used to meet in a small café near our pension, where Bing Crosby sang 'Don't fence me in' by the hour. Some young Spaniards were trying to get the English words, and eventually Robin wrote them down for them. Then the repertoire was varied with 'Three Caballeros'. We all learned the words of both songs very well.

Things were cheap but we had very little money left by this time. Most of us were selling spare clothes for a few pesetas. Whenever we were in funds after a sale we foregathered in the café and sipped anisette with our Spanish friends. On lean days we had ersatz coffee instead. The great thing was to sit there on the pavement and forget our worries for a while as part of the scene. We were accepted without reserve and it was good fun.

One morning Robin wanted to show me some beautiful glassware in a small shop nearby. We left Sheila, now blooming with health again, asleep in her drawer in the care of the proprietor of the pension and his wife. The shop was in a maze of narrow streets, one of many which sold delicate glass from tableware to jewellery in beautiful designs. There had been no tourists for years and the prices were unbelievably low.

Robin showed me a set of liqueur glasses and decanter and we coveted them for our future home. We had just sold a primus stove over the side of *Woodbine* and were in funds. We bought the set, feeling deliciously extravagant and wicked. It cost about £1 in English money. As we hurried back to Sheila Robin bought me a pair of earrings. They were irresistible, glass flowers with globes dangling from them, covered with tiny horns like mines. Clutching these absurd treasures and laughing we opened our door.

The drawer was empty. Sheila was gone. 'The Senora must have taken her,' Robin said, and we went down the narrow stairs. The Senora had not. She and her husband had peeped at her and she was sleeping. They had not disturbed her. They listened of course but had heard no sound. Suddenly we were alarmed. She was such a friendly child. Anyone could have taken her with a smile and a quiet word. We all rushed into the street – there was no sign of her. We ran to the café. There were a lot of people in that street and there, above their heads, we saw the blond curls. She was riding on the shoulders of a jovial Spaniard, having a wonderful time being paraded before an admiring crowd. We caught the word '*Inglisa*' several times as we hurried towards her. Her 'horsy' handed her to Robin with a beaming smile and a flood of Spanish. To this day we are not quite sure how she got there. We gathered she had been found in the open doorway of our room, and since everybody knew everybody else and all about the English baby from the trawler in the harbour, naturally whoever-it-was took her out to show everybody – because her parents weren't there, you see. Why then, not call the Senora? Well, she might not agree perhaps, so they had just gone quietly downstairs. Back we all went, Sheila and all, to Bing Crosby and the Three Caballeros – anisette all round. (No, not Sheila!) 'Don't fence me in,' sang Bing, and I thought: 'Yes, don't let us ever fence ourselves in,' as I looked at the people who in normal life we would never have known.

Corbeille's departure caused more speculation than ever. He had left some money for stores and said he would soon be back. Nobody believed him. A few more men left in passing ships for other destinations, but there was still no chance of one going south. On board *Woodbine* the remaining members of the crew ate through the stores, brought women and drink on board – and life in the Black Hole became even more unpleasant for Peggy and Noëlle. Mrs Hoppy seemed impervious to it all. She had seen too much in her seventy-odd years to worry about goings on in the fo'c'sle. It was different for the girls. After a few rowdy nights they hired a deaf and dumb boot-black. He sat outside the Black Hole all night with a cutlass across his knees. He could hear nothing, but felt every footfall as vibration of the deck. Robin asked the Consul to find out if the man was reliable and honest. Apparently he was well known and respectable. He had been blown up in the Civil War, the Consul said, and although his affliction was known to be functional, it appeared to be incurable. He would be a good watchman for the girls.

There was no news of Corbeille. The stores soon ran out and then the men started selling parts of the ship. Everything moveable went, and anything that could be unscrewed or taken apart. Buckets, ropes, canvas and fittings – all went ashore. Unless Corbeille returned soon there would be nothing but the bare hull left. The drunken Second Engineer had gone right over the edge, and was twice given clothes by the Mission to Seamen after he had sold what he wore. He even drank the alcohol out of the ship's compass – before somebody sold that too.

Woodbine was not the only ship in trouble at Las Palmas. A small wooden fishing boat from Estonia was anchored in the harbour, and we sometimes saw her people ashore. If we were an odd bunch in Woodbine, they were far odder. Her name was *Asje*, and she had been on her way to Argentina, packed with more people than she could carry stores for. They claimed to be Estonians but we noticed that they spoke

German to each other. Stores and money had run out and *Asje* had been in Las Palmas for months. There were some musicians aboard and they gave concerts ashore to make money, but they never got enough for the onward passage, so they ate and drank what they had in a series of parties. *Asje* was still there long after *Woodbine* had left on the final and most fantastic passage of her long life.

The authorities knew all about the 'Estonians'. They were escaping from the war crimes investigations in Germany, we were told. It seemed hard to believe. They seemed rather a slap-happy bunch of Cheerful Charlies whom it was hard to imagine as ruthless brutes. There was only one who fitted the part – a woman with hard, cold eyes and a bitter cruel mouth. The boys said she gave them the creeps.

Robin saw more of the Asjes than I did. Their skipper was a tall handsome man who had commanded a U-boat in the war. He was every inch a naval officer and I found it even harder to believe that he was alleged to be a war criminal – and yet, what on earth was a man of his calibre doing in an ill-found old fishing boat, trying to sail to Argentina? Robin chatted with him sometimes. They shared an interest in photography, and discussed cameras in a local shop.

Like his colleague in Madeira, the British Consul was doing his best to find passages in ships calling en route for Cape Town – or any other port in West Africa from whence there would be a chance of a coastal passage onwards – perhaps.

Meanwhile we seemed to have become part of the waterfront life of Las Palmas. Much of this flowed in and out of our friendly café where Bing Crosby fans fed coins into the jukebox and drank fizzy cold drinks with much talking and laughter. Sometimes the talking would stop and everybody concentrated on the drinks, their eyes turned away from the street. We soon realised that these sudden silences indicated the approach of the Guardia Civile. Walking usually in pairs, they would pass leisurely along the street, their faces impassive under their flat-fronted hats. It was our first experience of a police state. We wondered how it felt to be a young man who changed a jolly atmosphere in a café to one of palpable tension and fear by merely strolling past it. Did they enjoy the feeling of power? If you ignored the uniform they looked like ordinary peasants – po-faced, but that was no wonder.

We knew there were concentration camps on the island. The people had been on the losing side in the Civil War. The Anglican padre told Robin that there were British prisoners who had been in the camps for years and were by now barely human. He saw them sometimes but there was nothing he could do. They had been students who had gone to Spain to fight against Franco, full of high ideals, but mercenaries just the same in law. We were horrified to think of our own people among the prisoners existing without hope on this same small island. As the Padre said, it was a terrible price to pay for youthful ideals.

In the early evenings we used to sit in the café watching the ritual promenade of local boys and girls round the square. Freshly changed and dressed up the girls walked in groups round one way, and the boys, in their snappiest clothes, walked round the other. The groups eyed each other every time they passed but never spoke. It was a quaint old custom, and pleasant to watch while sipping anisette or black 'coffee' – according to the market in second-hand clothes.

Peggy and Noëlle spent most days with us, though they had to sleep aboard. Our room was tiny, with a single bed occupied by me and the new baby. Poor Robin slept on the ceramic tiles of the floor. One day Noëlle asked: 'How big is the baby now, Robin?'

'Oh,' said Robin, 'about as big as a haggis.' So poor Ann was known as Haggis from then on – until my mother put her foot down, months later, saying, 'It will stick to the child for life if you call it that once it's arrived.'

We seldom saw Mrs Hoppy. She spent most of her time on board since we had arrived at Las Palmas, but she took an interest in Haggis.

"Ave you got 'is things ready for 'im yet?' she asked me one day. 'Ought to be knitting for the pore little feller – not 'anging rahnd these 'ere forrin ports. It's not right.'

'Well, there's plenty of time really, Mrs Hoppy – nearly six months, and I have got some of Sheila's baby things still.'

The old lady snorted. 'Not even a vest for the pore thing,' she told Peggy and Noëlle, 'well, I'm going to knit one for 'im meself.' She had some knitting needles in her luggage, and I had some white wool in mine.

The girls gave me reports on the progress of the vest as Mrs Hoppy worked at it every day. I was really touched. It was the last thing that we would have imagined the tough old lady doing. Unfortunately there was not enough wool to finish the second side. Undeterred, she decided to sacrifice a knitted vest of her own. No need to waste time and effort, unravelling, and rolling the wool into balls. Sitting on her bunk in the Black Hole, wearing her vest, she unwound yarn from the bottom of it and knitted it directly into the little vest for the baby. The girls were fascinated.

'She passes it round behind her back to unravel it – and her vest is getting shorter as Haggis' gets longer.'

'It looks awful – all grey and much thicker than the wool you gave her. It's a pity because she did that very nicely, but if you only could see the poor old thing knitting away and unravelling as she goes.'

I could picture it, and funny as it was, I thought it was very kind of her to do it. We left before she finished it and Ann never got her vest, but I still think so.

ROBIN:

As soon as my family had recovered I took Joan along to the Consul's office with her passport, to be added to his list of British subjects. He had been trying without success to get us passages and the situation looked hopeless. Joan said:

'I refuse to have a Spanish baby, so I will come along and have it in your office.'

He smiled. 'Oh, I think we'll get you away before that.'

He did. I had been calling at the shipping agent's office daily for news of ships arriving en route to Cape Town. Then, one Friday, they told me that the Shaw Saville liner *Corinthic* was coming in on her maiden voyage, and was due on Monday. They had contacted the owners and inquired about passages for us. They had replied that although the ship was fully booked berths would be made available for us in the ship's hospital, subject to a fee of £65 being paid. The agents were quite distressed since I could not possibly raise money in England in that time. I thanked them and left disconsolately.

Then I had a good think. 'I am a ham. There's an eleven fifty-four transmitter aboard *Woodbine*. The batteries are flat – twenty-four volts of lorry-type accumulators. Ship's supply 110 volts D.C., supplied by the small steam engine near the top of the engine-room ladder. The D.C. generator would stand 20 amps easily – but the volts would have to be brought down to below thirty to charge those batteries.'

After these thoughts I hastily explained my intentions to Joan. Then I went down to the landing stage, where I met the boisterous crew of *Asje* about to go aboard their vessel, and accepted a lift to *Woodbine*. Their little dinghy was already full. When I gingerly stepped aboard and crouched down, holding both gunwales in between thwarts, crew and oarsmen, I realised that we had about ten centimetres of freeboard and any marked movement on board, or the arrival of the wash of a passing power boat would surely swamp us. They were quite unconcerned and we paddled gently away over to *Woodbine*. I got onto her rope ladder, taking care not to release too much weight on one side or other too fast. They departed singing some cheerful ditty.

On deck I spoke to one of the stokers who, with his pal, was still aboard, and explained the situation. They lit a fire under the boiler and raised enough steam to drive the small engine. I disconnected it from the ship's supply and with the engine revving slowly, put the bulldog clips onto the battery. The meter showed a slight charge. With the steam control I increased the charging current to twenty amps, and watched for a while. It charged away merrily all that afternoon and evening. The stokers put in the odd shovelful of coal down below, careful not to make any smoke.

The evening drew on and at 8 p.m. I took the batteries – heavy things – round to the corner of the 'wardroom' where my receiver and the 1154 were installed, and connected them up. I had already reinstated the ship's 110 volt supply to the little engine and opened its throttle so that the lights were working. I switched the receiver on to the 80 metre amateur band and did the same for the 1154. I switched to 'send' and pressed the key – hurrah! Current into the aerial! I listened round and heard one or two weak European signals – a bit early for 80 yet.

At 9.30 I heard two 'G's working each other in Morse code. Then a familiar fist came on – a ham I knew personally in Berwick. I called him and he came straight back, giving me full readability. I told him where I was and asked him to alert my friend Derek as I wanted to speak to him urgently. He said he had no telephone but would go down the road and use the public box there. I sat and listened. Within five minutes the familiar fist of my friend came on, calling me on the same channel. I told Derek what had happened (we all send and receive Morse as fast as one can write legibly) and asked him to ring my brother in Southampton. I asked him to say I needed £65 paid to Shaw Saville soonest, with the statement that it was for passages Las Palmas to Cape Town in *Corinthic*, and could he lend me the £65?

The reply came back. 'Your brother says you're an ass. The sale of your Austin Seven fetched £120 and he will get £65 to Shaw Saville agents soonest.' After thanking my friends in England and the two stokers, I went ashore and told Joan to hold thumbs.

On Monday morning I went to the agent's office. I was greeted by the pretty young clerk there who said, 'Doctor, we cannot understand it. Shaw Saville have received £65 and told us to issue two tickets to Cape Town. The cable office has been closed over the weekend. How did this happen?'

I grinned, and said, 'It's all done by mirrors.' She shook her head. I thanked her and left with the tickets.

Next morning, in full daylight, the agents had our luggage hauled out of *Woodbine's* filthy hold and dumped in a very small lighter. *Corinthic* was alongside the quay, and there was a large flat pontoon on her seaward side. Our luggage was to be hauled on board from there.

Together with the agent's stevedores and Sheila, Joan and I sat on the boxes anxiously watching the diminishing freeboard. I was not going to leave my magnificent *fogos d'artificios* in *Woodbine* to be sold by the crew, so I had cut the sticks off, put the heads into one of the boxes – and forgotten about them. Arrived at the pontoon the stevedores unloaded the boxes onto it and Joan took Sheila up a gangway. Then, while an officer and seamen were preparing to haul the boxes on board, the stevedores departed with the lighter. A cargo hook had been lowered but there was nobody to hook our boxes onto it.

I was down on the pontoon in my grey going-aboard suit. 'All right,' I shouted. 'I'll do it.'

One after another I arranged the sling round each item and anxiously watched it swing skyward and inboard. When the last one had gone up I went up the gangway and joined Joan, Peggy, Noëlle and Sheila.

JOAN:

When Robin's machinations and the Consul's backing suddenly achieved our passages in *Corinthic* we were very anxious about leaving Peggy and Noëlle still stranded in a foreign port. Las Palmas was very different from friendly and hospitable Funchal. The nursing sisters were right – we never saw a British resident all the time we were there. We had to leave the girls still living aboard *Woodbine*, short of provisions, with a disgruntled crew and a shell-shocked Spanish watchman. We knew the Consul and the British padre would do all they could for them, but it was horrible to break up our tight-knit foursome, with Robin's broad shoulders to lean on.

Boarding a beautiful liner, immaculate on her maiden voyage, with fourteen wooden boxes, smoke-grimed and stained with sea water, assorted suitcases, a carry-cot and a toddler, under the fascinated eyes of her smart passengers, calls for a certain aplomb. They had been lining the rails watching our slow progress across the harbour in the overloaded lighter, and now they watched Robin hitching the cargo hooks onto the boxes as the ship winched them on board. I was beyond caring what impression we were making, but when, at the top of the gangway I heard a loud male voice say, 'Personally, Ay think people who get into trouble in foreign ports should be sent straight home,' I did feel angry.

With our disgraceful luggage decently out of sight we were taken to the ship's hospital. It was large and comfortable and I wished we could stow Peggy and Noëlle away in it with us. Soon we were all taken to the empty dining saloon for a belated breakfast.

How we fell upon the bacon and eggs, English marmalade, and real coffee! The amused steward brought us second helpings each without being asked. Then, as we were finishing this truly British feast, we became aware of another ship moving slowly past our portholes – one of which suddenly filled with a red ensign. The stewards must have thought they'd shipped some real lunatics as we leaped up and rushed for the deck. Perhaps if that ship was going to South Africa she would take Peggy and Noëlle . . .

When she had settled into the next berth, Robin and the girls went aboard her and told the Captain their story.

He was sympathetic but regretted there was nothing he could do. The Consul had already contacted him. He was horrified at the way the girls were living and would have taken them if he could. Apparently the owners had refused.

So, our hopes dashed again, we had to say goodbye. Once more at sea 'on passage towards Cape Town', I promptly went down with sandfly fever.

I was not seasick in *Corinthic*. I forget how long the fever took to run its course. I was already in the hospital together with Sheila and Robin. I remember Sheila crawling out of the door, wearing her upside-down carry-cot like the shell of a tortoise. She had got to the exhausting mobile stage and Robin had to cope with it alone.

ROBIN:

One morning when I was washing nappies in the passengers' laundry a tall, good-looking woman with blond hair brushed up into an Edwardian crown was washing things in the next basin. She smiled at me and said, 'Don't use that soap. I've got something much better,' She went to her cabin and fetched a small cardboard packet of white powder. 'It's not a soap,' she explained, 'it's detergent.' She and her husband were going to South Africa to promote the new product. It worked beautifully. I didn't know then that the white powder was to revolutionise washing in laundries, households, and factories, and the yellow soap would almost disappear from our lives. I was most grateful to her and we met and chatted often over our daily chore.

Although we were berthed in the hospital we were treated as ordinary passengers by the ship's people, and, once it was known that we were not undesirable deportees, by most of the passengers too. Naturally our unorthodox arrival on board had caused quite a stir of interest and several people asked me for our story. None were as smug and unpleasant as the man whose arrogant remark had annoyed Joan at the gangway.

Woodbine and her people must have been a topic of conversation among the passengers, for later I was approached by a man who was (unbeknownst to me) a famous film producer. He had just finished making a film in Australia. He asked me to meet him in the lounge that evening after dinner and tell him all about the voyage of *Woodbine*. I told him everything. He listened silently. When I had finished, he said, 'Write it down as a story and send it to me.' I said I would consider it seriously and he thanked me. Knowing what I now know I wish I had taken him up on this instead of writing it forty-seven years too late!

The ship's officers showed great interest in the story, particularly the radio officer. When we were a few days out of Cape Town, he told me that H.M.S. *Vanguard* was not far away, on her way back to England with the Royal family aboard after their tour of South Africa. The Captain of *Corinthic* had given him a loyal message of goodwill to send to *Vanguard*. Unfortunately he was unable to send it because there was a fault in the ship's transmitter. I asked him if I could have a look at it with him and he agreed. We opened it up together and traced the fault to the power pack supplying power to the final amplifier driving stage. I suggested that we disconnect this unit and switch over to another which was lightly loaded and this we did. It worked! The message was sent and acknowledged.

8. 'BETTER LATE THAN . . .'

JOAN:

Three months after leaving Southampton, and in a very different ship, we reached Cape Town. I had more or less recovered from the fever and Robin called me on deck to see Table Mountain as we approached. It was not the first time either of us had sailed into that harbour from the sea (nor was it to be the last!) but it was a very different arrival for us. Our little family was News.

We had hardly been seen by the immigration officers – still on board – before we were pounced on by reporters. Unflattering photographs and various versions of our adventures appeared in the Cape papers which were duly sent to my family by friends.

The grotty luggage passed through customs without any trouble, including Robin's genuinely forgotten 'fogos'. I think they would have been frowned upon even then – in these days we'd probably land in gaol. The boxes were handed over to South African Railways and Harbours to be sent on to Natal.

We were met by old family friends of my parents, who collected us from the ship and took us to their lovely home. Robin had never seen ridgeback dogs before and when three huge friendly animals came bounding towards us, with, he thought, their hackles up, he grabbed Sheila, convinced they were about to attack her. Our stay was very pleasant and gave me time to pick up after the fever before we were to take the long train journey up-country.

There was no news of Peggy and Noëlle, and we wondered what was happening to *Woodbine* – the Skipper, the old doctor, the Hoppys, the two stokers, the kind-hearted Chief and the rest of our shipmates still with her.

The train journey took two days and a half and two nights. I was disappointed that the train passed the Hex River Valley in the dark. As children my sister and I used to watch out of the window, sometimes getting smuts in our eyes, to see the steam engines – always two – as the train wound round sharp bends on steep gradients. I had described the miles of flowers in the Karoo, if one were lucky enough to see it after rain. Robin was not so lucky but he was fascinated with the railway engineering and the big Garrett steam engines.

On the second day we came down from the Drakensberg through Van Reenen's Pass at dawn. It was about eleven o'clock when we looked down on Pietermaritzburg in its hollow. About an hour later we pulled into the station – and there was Mother, running alongside the slowing train as Robin held up Sheila to show her.

There had been changes to the home I had left three years before in a war-time convoy. Mother and Father had a wing built on to what had been a small house and they had moved into that, leaving their old rooms ready for us, my sister and brother. They were delighted with their first grandchild and were informed that they were to have another.

We had caused them a great deal of anxiety. My sister had heard on the radio early one morning that *Woodbine* was reported missing and had broken the news to the family. Headlines had appeared in the Natal papers: 'Local girl missing at sea' and so on, with bits about my modest military service and our Cairo wedding. Now there were more interviews and photographs – but the journalists were mostly locals too and seemed genuinely pleased about the happy ending. The subsequent articles were more accurate than the others and free from embarrassing journalese embellishments. (Some of the stories in the overseas press were quite awful! One had the four women sailing from Madeira 'weeping copiously'.)

Adventures are really only fun in retrospect. We would have more in the future but none quite as colourful as the ones we shared with our shipmates in *Woodbine*. Now I was very happy to sit beside Sheila's playpen under the jacaranda trees being blissfully spoiled by my family. At last I was looking like an expectant mum and not a half-starved survivor.

Ann was born with the touch of drama that had surrounded her from the first. Robin took me to the hospital in the small hours and we tried not to wake up the household. He free-wheeled our ten-year-old car down the steep drive before starting it up. Father *had* woken up. Thinking the car was being stolen, he dashed out with his automatic pistol at the ready. Luckily he remembered the imminent arrival of his second grandchild before blasting our tyres.

At last we had news from Peggy and Noëlle – and this is where we hand over to Peggy to tell their side of the story.

Pss-kerjoink!

MY STORY

BY PEGGY

Noëlle and I first met when we were 16-year-old students at Pitts Cottage in Westerham, Kent. We'd been accepted to learn catering at Pitts Cottage, which was a well known tea-house. It was also famous because years ago it had been William Pitt's shooting lodge. Winston Churchill and his family lived in Westerham at 'Chartwell'; this was their country home and little Mary Churchill's nanny sometimes brought her to have afternoon tea at Pitts.

Noëlle's father had been a dental surgeon in Dublin and consequently she spent her childhood in Ireland. During the war she had joined the F.A.N.Y.s (First Aid Nursing Yeomanry). The name originated during World War I when a band of women offered their services as ambulance drivers and nurses. During World War II it became a very much larger organisation with girls undertaking a variety of jobs, so much so that several distinguished themselves for various acts of bravery.

Noëlle was sent by flying boat to India where she worked with agents in Burma. She was a real tomboy and always great fun. I well remember hearing how some of her friends had arrived at a party leading an elephant with Noëlle up top.

I will not go into my history here except to mention my family owned an hotel in Caterham, Surrey. During the Battle of Britain the hotel was burned down and my mother and six of our guests lost their lives.

Later, after a spell in the American Red Cross, I applied to join the

F.A.N.Y.s and was sent to Bedford on a preliminary course but the war ended before I was enrolled. As a result we were allowed to stay at the F.A.N.Y. Hostel in London until we found other jobs. We were all keen to go overseas and I shortly joined the Control Commission in Germany. Most of us were sent to Minden in Westfalia. I would like to tell of my experiences there and the many friends I made and worked with, but that is another story. Much of the time spent in Germany did influence my future life and looking back, 'It's destiny that shapes one's ends.'

I do consider myself very lucky to have worked and thoroughly enjoyed my two years in the Control Commission, but, as usual, all good things have to come to an end. There were rumours that our R.D.R. Division (which was responsible for restoration and returning works of art to their rightful owners) had served its purpose and a number of us would be made redundant. It was a depressing thought and that Christmas (1946) we tried to dismiss it from our minds. Then completely out of the blue the following telegram arrived for me: 'Are you willing pay £100 fare to South Africa – sign on as crew – trawler sailing last week January. Good fun. Reply urgent. Noëlle.' It transpired Noëlle had arrived back in the U.K. from India and had spotted a notice in the Personal Column of the London *Times* advertising this adventure. Noëlle's mother had gone with her to meet the advertiser, and after checking various channels they concluded this Mr Corbeille had made two previous trips to South Africa. Apparently he was a South African and was involved in bringing goods to the U.K. (and then taking English people who were prepared to act as paying crew to South Africa). There were to be about twenty-six people all told and dancing on the deck was mentioned. During the war Service personnel had often spoken about calling into Durban and the wonderful hospitality they had received there. The Stardust night club also sounded rather special. The answer had to be 'Yes' – we'd both sign on. It was a mad rush leaving Germany.

I stayed with Noëlle and her mother in Beckenham, Kent, and on arrival we received a card from Corbeille dated January 28 1947 instructing us to 'be aboard Woodbine on Saturday January 30 1947' as he hoped to sail early the following week. It said 'bring for your own convenience sheets, pillows, blankets and deck chairs, also reading matter. Please confirm by return post. Corbeille. P.S. Please note change of plans, come aboard at Southampton. Woodbine, Town Quay, Southampton.'

England was experiencing a freezing cold winter and everywhere was covered in deep snow. We said our farewells to our families and friends and most people said they were envious of our departure to warmer climates.

The train took us, plus our twelve pieces of luggage and two big tin trunks, down to Southampton. We asked the taxi driver to take us to Woodbine's berth at Town Quay. He said he knew it but didn't enlighten us until he pointed out a little funnel sticking up above the quayside. Sure enough, it was Woodbine, and there was a man standing on the deck. He turned out to be Robin, who became our 'big brother' and lifelong friend.

'Could you please tell us where we can find the gangway?' we asked.

'There isn't one,' replied Robin. 'Are you the two girls who are coming aboard? This ladder acts as the gangway and if you can step on the rungs I will hold your hand to help you balance across.'

How our luggage got on board I will never know. The ship was far from ship-shape and we were informed she weighed 308 tons, and in the past had carried water to the troop ships around the area. Our meeting with the ship's company is now rather vague, especially as some people had decided to remain on shore and would hope to reclaim a refund on their fares!

'Down under' was a pokey little type of wardroom amidships, where we met Captain Corbeille and his bride. They both wore what looked like naval uniform, fictitious as we soon discovered. Robin, his wife Joan and Sheila, their 14-month-old daughter, were going to Pietermaritzburg, Natal, where Joan's family lived.

Pat and Bill were a newly married young English couple, and this was to be their honeymoon trip. Pat took one look at the accommodation and decided to rush back home, but Bill opted to continue. Pat eventually did manage to get a passage in one of the liners going to Cape Town and was prepared to await her Bill's arrival.

There was a little Major, a retired army medical officer without ties, and an ambition to travel the world in his own time. Mr & Mrs Hoppy, a down-to-earth cockney couple who had met in a London pub and had recently married. He had served in the Merchant Navy but was not very steady on his feet, having been shot through an ankle during the war. This would be the first time his new wife had been to sea. She had little hair and always wore a scarf tied like a turban. The purpose of their trip was that Hoppy's first wife and family had left him before the war and he had discovered they were now living in Rhodesia. He wished to see them again and wanted to meet his grown-up children. The Hoppys were paying their own fares and Hoppy was to act as Second Mate.

Noëlle, Mrs Hoppy and I were installed in a so-called cabin in the fo'c'sle. It consisted of three bunks and a small chest of drawers. As stewardesses we were expected to sit on deck and peel the spuds in buckets of freezing cold water, and would later be required to help with the catering. As there wasn't any definite date for sailing and nothing was organised, Noëlle and I decided to book into the Dolphin Hotel for a night or so. Then after returning to our cabin we went to the public baths in order to have a decent shower. We rang home and explained that the sailing date had been delayed, all was well, etc., but we didn't divulge that we were keeping our fingers crossed for improvements!

Joan and Robin have described the alarms and excursions of the last few weeks, and the eventual departure from Southampton.

I remember going down to the engine room and being violently sick from the smell of hot oil. I do recall Noëlle stuffing me with lots of dry bread which she said would help me get my sea-legs. At first the weather was not too bad but when we were about eight days out a great big storm blew up, waves over the deck and we had to crawl on our tummies to get midships from the fo'c'sle. The ship tipped to 40° and we were told later that, had she tipped to 45°, we'd have been in Davy Jones' locker.

One day (before the storm) Noëlle and I decided to go into the galley to see if we could find enough fuel to make some rock cakes for everyone. We had always been taught to sift the flour but of course under present circumstances this was impossible. Imagine the shock when we found some false teeth in the flour! The Dutch cook had dumped his teeth before rushing off to be violently ill.

When we ran out of coal we simply drifted and hoped that Robin could keep in contact with Madeira. His radio batteries were getting low and I can recall Noëlle and me helping to prop him up as he was trying to keep his balance whilst waiting for his schedules with the operator there.

As we tossed about without coal an American Liberty ship slipped by within a mile or so, and the boys sent up our one and only flare, but unfortunately no one spotted us. Later we heard that the B.B.C. news had announced that *Woodbine* was missing with 24 people on board, and reporters called on my brother and his wife asking: 'Did you have a sister Peggy on board the missing yacht *Woodbine*?' Panic all round at home.

Things became really dangerous, and again, Robin tells the story first-hand.

We arrived in Madeira looking like tramps – we had been unable to wash, not to mention our starvation diet. Madeira gave us a great reception with crowds of people and some reporters coming on board. It was all quite overpowering. It must have been quite an event for the locals because post-war visitors to the island were few and far between due to sterling restrictions and the non-availability of passenger ships. The porter from the Savoy Hotel did make himself known to us and explained that due to these reasons the hotel had only three visitors, and should any of us require accommodation the owner, Mr Carlos Dios, would be happy for us to be his guests; the charge would be only the equivalent of 10/- a day in escudos. Several of us were lucky enough to have travellers' cheques so we offered our appreciation and promised to let him know later. The next man to come forward introduced himself as Costa Gonzalas to Noëlle and explained that he was the Greek friend of a French girl called Angelle. He used to know her in India and he felt sure he'd met Noëlle in the F.A.N.Y. hostel there. This happened to be quite true. He told Noëlle he was the Captain of the *Jeannerie*, a beautiful yacht which was anchored near our old tub. She had been built for a wealthy Greek and he was delivering her to the new owner in the States. He had a crew of 8 cockneys aboard and anticipated a stay in Madeira for approximately another week. We both almost threw our arms around him when he invited us to go aboard and said we would be welcome to use the facilities such as hot water, baths etc. At the first opportunity and within a few hours our dreams of life in a real yacht finally came true! The crew were so courteous, we had the best baths ever, and afterwards we were shown around. What luxury! The main bedroom was like a film star's and I can remember a gold dressing table set fitting into individual recesses of the built-in vanity unit. The wardroom had special panels that when opened, revealed a cocktail cabinet, radiogram etc. We returned to the old tub as civilised human beings! Robin, Joan and Sheila had been taken to their home by the Reuters' agent and his wife. Together with the Major and the Hoppys we decided to spend some of our money and accept the hotel owner's kind offer to stay in his hotel. What luxury! It was a first-class hotel overlooking the harbour of Funchal, with terraced gardens going down to a swimming pool at the water's edge. Bougainvillaea was in flower everywhere. I well remember the lovely perfume of some of the flowers that were new to me. We were offered excellent service by the few staff who were employed, due to there being only three resident guests, namely the British Consul, his wife, and a wealthy Englishman from the North of England who was there for medical reasons. We were all thankful to have a good meal and lovely beds.

Costa appeared, to enquire if we had settled in, and introduced us to Tom Lee, a Welshman who lived on the island and was employed by Blandy Bros., the agents responsible for victualling visiting ships. Social life took a new twist as Noëlle and I spent most evenings being shown round the local night spots, returning to the hotel for breakfast in bed. Our first outing with Tom and Costa was on St David's Day, which was a good excuse for a Welshman to let his hair down and for us to join in. Tom had an amusing way of composing little poems about the various events that had happened to us during the day.

The British Consul and his wife invited us, now so-called hotel residents, to an official reception organised by the Portuguese for the officers of H.M.S. *Duke of York* which was paying a goodwill visit to the island. We girls had been offered a ride in a bullock cart the afternoon of the party and had forgotten to pay special attention to the time, so imagine our embarrassment when we were driven up to the hotel in this mode of transport as the other guests were beginning to arrive. After a quick change we regained our composure and thoroughly enjoyed the evening. The ballroom certainly lent itself to such an occasion and the floral arrangements were exquisite.

Another day, whilst enjoying one of the hotel's delicious luncheons, the idea was hatched that it might be a nice gesture to invite some of the not yet paid crew round for dinner, because, after all, they'd done us no harm and had always been courteous, and they might appreciate a little outing. Arrangements were made and a good time seemed to be had by all. The following morning we were greeted by a very embarrassed Carlos Dios, who said he hoped our visitors had enjoyed their dinner, but unfortunately the waiter had reported several cruets were missing. Our embarrassment was even greater than Mr Dios' and we hurried down to the ship to retrieve the missing articles. No grudges were held against us because the following Christmas we received a card from Mr Dios saying he sent us greetings and hoped to see us again in the not-too-distant future.

After two weeks on 'cloud nine' the bomb fell – a message from Corbeille ordering us all back on board as *Woodbine* was due to sail to Las Palmas. There followed the comedies described by Joan, with *Woodbine* under arrest!

We had been in Madeira for six weeks and all that time Robin had been closely involved in the machinations and dramas he and Joan so vividly remember. When the owners of *Fort St Joseph* generously released *Woodbine* from arrest, Corbeille's peremptory order, flung over his shoulder on the street, had us all aboard again for the passage to Las Palmas.

Our friends on the island came aboard to wish us 'bon voyage'. I well remember my glasses of Madeira wine and those of apricot brandy. The latter I have never touched since, because, with Funchal just out of sight, we hit a beam sea . . . and the rest I leave to the imagination.

Three days later we arrived at Las Palmas – in an atmosphere of understandable mutiny. The crew had not been paid since leaving the U.K. Corbeille said he was under financial pressure and would be flying back to England. It was essential that he should arrange finance for the rest of the voyage, and, having done so, he would return to Las Palmas. As we were anchored in the harbour arrangements were made for a launch to deliver essential stores and goat meat.

The smouldering resentment flared up into a real mutiny, and after the dramatic scene in the Consul's office Corbeille left the island. Joan had been admitted to the local seamen's hospital and Robin had found a room for himself and Sheila in a pension near the docks. They were having a very bad time at this stage and missed the next scenes aboard *Woodbine*.

After Corbeille's angry departure the paid crew became very busy organising an auction sale of ship's belongings – chronometer, etc. The sale took place on deck and a very questionable crowd of Spanish bargain-hunters arrived. I do remember we were left with quite an odd assortment of drinking utensils, photographed for my album.

Chief and company were now able to order themselves locally made suits, and would be saying fond farewells to us as they intended joining other ships. Chief, the good-hearted Irishman, kindly offered us what we felt would be classed as 'hush money' from the sale, but we refused, saying we had not seen what took place on deck. He then offered the Hoppys and us girls a day's taxi trip round the island, so we thoroughly enjoyed the outing with him, seeing the vineyards, tasting wine, etc.

The ship's company dwindled down to the Major, the Hoppys, Monty, Bill, Noëlle and myself, and somehow we acquired a local to cook for us in exchange for a meal.

Joan and Robin have described their two weeks in the pension after nearly losing both Sheila and the new baby, and the life on the waterfront. Only he could tell of the struggle to get onward passages. His radio operating had saved us before Madeira, and now his ingenuity and radio skill finally got the passages in *Corinthic* – but only for him and his family.

We were green with envy when we went aboard to see them off. They were installed in the ship's very clinical hospital, which they had to themselves. I remember Joan saying that we could easily stow away in their accommodation, but an officer appeared and invited us to stay aboard for breakfast.

All efforts to get passages for Noëlle and me had so far failed. After Robin and Joan sailed in the *Corinthic* the padre from the seamen's mission and the British Consul were keeping a fatherly eye on the two of us. They kindly invited us ashore for the odd meal and on one occasion they had taken us to a cock fight! Most of the days were spent sitting on deck because we didn't have pesetas; the radio had been auctioned and we were heartily sick of eating old goat. The saving grace had been the Major, who did have a few pesetas which enabled him to get ashore to catch the weekly airmail and also to bring back a local newspaper. Luckily he had a Spanish/English dictionary and so was able to act as interpreter. This became the evening pastime until the oil lamp ran out of fuel due to the fact that the 'no-hope alcoholic sailor' had drunk the methylated spirits. Poor old Mrs Hoppy had almost given up, and wasn't even bothering to dress or get out of her bunk. The baby's vest she was knitting was unfinished when Joan left, but she sat in her bunk continuing to 'add on' straight from the bottom of her own very off-white garment.

Peace on deck was suddenly shattered when Noëlle came rushing along saying: 'He's back'! Sure enough, there was the dreaded Corbeille still in uniform, striding towards us saying, 'Noëlle, Peggy, what is all this? What have you to say?' We replied that we were still stuck here through no fault of ours, and we were not responsible for anything. To our great relief the Padre and the Consul appeared and announced that 'the girls will be leaving the ship and arrangements have been made for them to be taken to the seamen's mission.' A launch appeared and our 12 pieces of luggage, together with the now very rusty tin trunks, were resurrected from the hold.

I'll never forget the sight of a donkey cart with a very bony little donkey waiting at the quayside to transport our possessions to the seamen's mission. The mountain of luggage was loaded onto the cart and Noëlle and I tramped along behind. The main thoroughfare was quite a sandy type of road with funny old cars and the odd open bus honking their horns. In the midst of all this the worst happened! The cart wheels doubled under, the donkey fell down, and there was just a big heap in the middle of the road – if only I had had my camera! An S.O.S. went out to the seamen's mission and all hands on deck quickly installed us and our goods into the mission.

The Padre had given up his bedroom for us and I can remember the big mosquito net which made the bed look like an old-fashioned draped four-poster. The bedroom door opened onto the beach and the sand was really golden. Things were now moving fast because the following day the Padre told us the *City of Exeter* was due to arrive that afternoon, and the Foreign Office had arranged for Captain Andrews to come and see us. As promised Captain Andrews together with the British Consul appeared and informed us that, provided all our papers were in order, he was prepared to take us and we would be listed as Distressed British Seamen. He hoped we wouldn't mind having to sleep in the children's nursery at night as all the cabins were full. We requested that please could we go on board very early in the morning because we were embarrassed about our luggage and didn't want to embark in front of the passengers. This request was granted, but Noëlle's Irish paddy was secretly ruffled when a tall young Dutchman standing at the top of the gangway announced: 'You girls are new around here, aren't you?' Later, Peter Dixon became a very good friend of Noëlle's. The crew of the *City of Exeter* pulled our legs and informed us that 'all distressed British seamen have to do the donkey-work on ships', and that there would be no shortage of jobs for us.

Captain Andrews had instructed that we were to mess with the officers and were to be treated as passengers. Our passages were paid for by the Foreign Office and later the bill would be presented to Captain

Corbeille!

Life on the *City of Exeter* suited us very nicely, thank you, and we enjoyed meeting the passengers, the crossing the line celebrations and generally taking part in the various shipboard activities . . . and, of course, dancing on the deck.

We finally spotted Cape Town and Table Mountain, and the following day we passed immigration and our passports were stamped. We collected an accumulation of mail and booked into the Overseas League of which we were members. The next two days enabled us to catch up with returning to normal life ashore, and also to see a little of Cape Town. On June the 18th we boarded the train for a three-day journey to Durban; the cost of the ticket was £7:6:0d! It was on this journey that we received a telegram from Joan's parents telling us to leave the train at Pietermaritzburg as they wanted us to stay with them. The whole family were at the station to meet us. We met a number of Joan's relations and (at a later visit) we were introduced to tiny baby Ann. We felt as if we had come home and I don't think we stopped talking during our ten-day stay there.

After some further travels, in Rhodesia etc., we reported to the Majestic Hotel in Durban as instructed, to work as receptionists. We were allocated a pleasant bedroom and were treated well by our new boss and the head receptionist. It was a matter of weeks before a couple of familiar faces appeared at the reception counter. The Major had read that the *City of Exeter* was berthed at Durban, and had introduced himself to Captain Andrews, hoping that he might know our whereabouts. Together they had decided to pay us a surprise visit and they waited until we came off duty to take us out. I must admit Noëlle and I later had a giggle when we pictured ourselves doing the town with these two little men who were so much older and shorter than ourselves. We were able to catch up on events that unfolded after we left Las Palmas because the Majors, the Hoppys, Bill and the new crew with Corbeille in charge, had stuck it out to the bitter end. We heard of their risky voyage, minus the auctioned instruments, the chronometer, etc., and how, after reaching the coast of West Africa, Corbeille had announced they were coming into Freetown which actually turned out to be Lagos!

The Major told how the Hoppys had packed their overnight bag in preparation for going ashore next morning. They'd hung the bag in their open porthole, little realising that the locals come along with long hooks and steal goods through open portholes. The worst happened and so it was goodbye to Mrs Hoppy's bag and also her glass eye! She was devastated and it was decided that they should report the theft. Having set off along the sandy beach a big African was spotted staring closely at something in his hand. Mrs Hoppy went straight up to inspect, grabbed the object and gulped: 'That's me eye, mate.' The Hoppys had learnt their lesson the hard way, but at least she did get her eye back!

From Lagos they finally made it into Rhodesia and they did locate Hoppy's family, but, owing to a very cold reception, there was no option but for them to return to England as soon as possible. Thus they were able to contact us in Durban before joining the ship that was to take them to the U.K.

Bill eventually reached Cape Town and found his bride waiting for him there. They also made their way to Durban and when last we saw them they had jobs and intended settling locally to live happily ever after.

I never heard what finally happened to Captain Corbeille. He was certainly a force to be reckoned with, and I wonder if nature's elements mastered him in the end? As for his ship, could she have met the same fate as her master?

EPILOGUE

BY JOAN

None of us ever heard what eventually happened to Corbeille. We got the last news of *Woodbine* several months after we left her, when we had acquired a small cottage near the family house.

One day two tough-looking characters turned up at our door – our helpful and very efficient stokers from the ship. They had traced us to our new address.

Sheila was delighted to see her old shipmates. They duly produced chocolate for her and were introduced to baby Ann who was in her cot near the garden wall. Life was rather dull just then after all we had been through together, and soon we were swopping news and reminiscences. What the neighbours thought of loud voices and somewhat lurid nautical language on our side of the wall, Robin and I could only imagine.

The account of the voyage from Las Palmas lost nothing in the telling. No need for the proverbial 'pinch of salt' – we had already experienced Corbeille's methods. Without proper instruments, no chronometer, and a largely scratch crew of wharf-rats, he excelled himself on that passage. Somehow they made two ports of call – Bathurst in the Gambia and Freetown in Sierra Leone, where they took on some coal and supplies. Ten days out of Freetown, inevitable disaster struck.

When the professional stokers came on watch they saw with horror that the boiler was dry and the crown sagging down on the fire.

'We never drew a fire quicker in our lives,' they told us. 'A few more minutes and *Woodbine* would have been blown to pieces.'

Once again Corbeille's famous blanket sails were rigged and again proved useless. After drifting for another ten days, without lights, they were found off Accra by a Swiss ship and towed into Lagos. *Woodbine* was arrested for the second time, and the last the stokers had heard was that she was to be sold for salvage.

As cousin Tom had said in Southampton, she was a good little ship.

ADDENDUM

BY BARBARA

Back at the ranch, or in Pietermaritzburg if you prefer it, Joan's family were only too well aware of what was going on. When we first knew that Joan and Robin were coming out in a yacht Mother and I happened to be in Durban, so we thought we would try to find out what sailing in a trawler yacht would be like. In those days there was a similar-sized vessel that took folks for a trip roundabout the seas outside the harbour. We joined her for a morning sail. This was a very noble act on my mother's part, for she was the world's worst sailor. So I sat her down in a comfortable chair on the deck, and then, as we passed the Point, I sat on the deck myself and was quite unable to get up again until the return to the quiet waters of the harbour. I admit I was laughing rather hard as well as trying to rise.

Thus we did know a little of what was in store for my sister and her husband (whom we had not yet met). None of us lacked imagination, but Dad said: 'You can't sink a trawler.' Of course he did not know about Corbeille and the house coal. The Queen Mary could have sunk if she had run out of power.

Very early one morning a kind neighbour came to me and said, 'I have just heard the six o'clock news. Your sister's boat is in trouble, so tell your parents before they hear the seven o'clock news.'

It was as well that we had that warning to cushion the shock a little, for the bulletin told us that *Woodbine* was drifting in a full gale off

Madeira, that she had run out of coal, and that there was a pregnant woman on board (they did not say a hospital case). Because we knew that Joan was the only young married woman on board, and because of the much stricter morals of the time, we knew it must be Joan.

Dad spent a fortune in cables that day, to the Port Authorities in Funchal, the British Consul and anyone else he could think of. We went through several days of intense anxiety until we heard from Robin and saw that press photograph, which, blotchy as it was, at least showed us Joan standing up, and Robin with little Sheila in his arms.